Minutes a Day-Mastery for a Lifetime!

Level 3

Mathematics

2nd Edition

Nancy McGraw

Bright Ideas Press, LLC
Cleveland, OH

Summer Solutions Level 3

2nd Edition

Printed in the United States of America

ISBN-13: 978-1-934210-35-2
ISBN-10: 1-934210-35-8

Cover Design: Dan Mazzola
Editor: Kimberly A. Dambrogio

Cleveland, Ohio

Instructions for Parents/Guardians

- *Summer Solutions* is an extension of the *Simple Solutions* approach being used by thousands of children in schools across the United States.
- The 30 Lessons included in each workbook are meant to review and reinforce the skills learned in the grade level just completed.
- The program is designed to be used 3 days per week for 10 weeks to ensure retention.
- Completing the book all at one time defeats the purpose of sustained practice over the summer break.
- Each book contains answers for each lesson.
- Each book also contains a "Who Knows?" drill and *Help Pages* which list vocabulary, solved examples, formulas, and measurement conversions.
- Lessons should be checked immediately for optimal feedback. Items that were difficult for students or done incorrectly should be resolved to ensure mastery.
- Adjust the use of the book to fit vacations. More lessons may have to be completed during the weeks before or following a family vacation.

Summer Solutions Level 3

Reviewed Skills include

- Addition and Subtraction With Regrouping
- Place Value
- Rounding to Hundred Thousands
- Basic Multiplication and Division Facts
- Multiplication of 2- and 3-Digits by 1-Digit
- Division of 2- and 3-Digits by 1-Digit with and without Remainders
- Basic Geometry
- Simple Fractions and Decimals
- Charts and Graphs
- Math Vocabulary
- Telling Time and Finding Elapsed Time
- Counting Money
- Simple Measurement Conversions
- Word Problems

Help Pages begin on page 63.

Answers to Lessons begin on page 75.

Lesson #1

1. $325 \times 5 = ?$

2. How many quarts are in 3 gallons?

3. $65 \div 5 = ?$

4. A quadrilateral has ________ sides.

5. $628 - 199 = ?$

6. Round *654* to the nearest hundred.

7. $6 \times 6 = ?$

8. Jason spent $4.65 on a book. David spent $3.79 for his book. How much did the boys spend altogether at the bookstore?

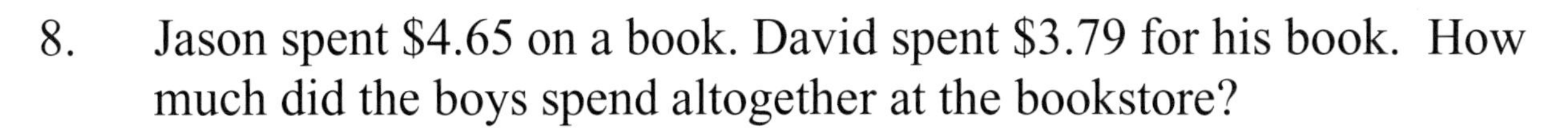

9. $187 + 469 = ?$

10. Paul has 2 quarters, 2 dimes and 1 nickel. How much money does he have?

11. Which digit is in the thousands place in *36,349*?

12. Write *4,216* in expanded form.

13. If it is 7:15 now, what time was it 3 hours and 10 minutes ago?

14. A six-sided polygon is called a(n) ___________.

15. $846 \div 2 = ?$

16. Is *5,217* an even number or an odd number?

17. 9,346 ◯ 9,463

18. Draw a rectangle and shade $\frac{3}{4}$ of it.

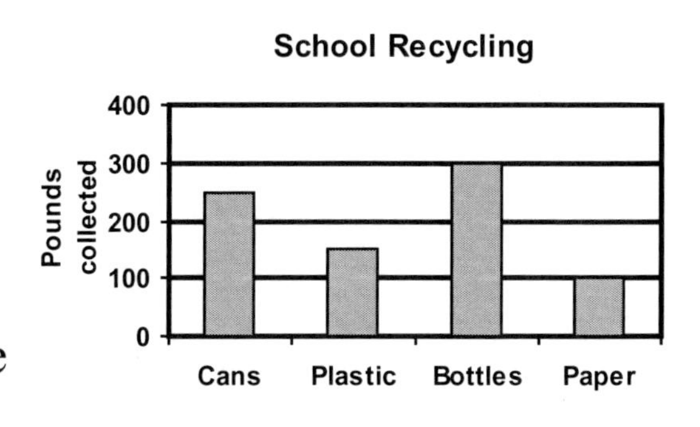

19. How many pounds of paper did the class collect?

20. What weight of cans and bottles were collected?

1.	2.	3.	4.
5.	6.	7.	8.
9.	10.	11.	12.
13.	14.	15.	16.
17.	18.	19.	20.

Lesson #2

1. Find the mode of this set of numbers. 27, 34, 19, 27, 39

2. $50 \div 4 = ?$

3. A pentagon has ____________ sides.

4. Write *3,000 + 600 + 20 + 7* in standard form.

5. Draw intersecting lines.

6. Jared has 56 baseball cards to save in a book. If 7 cards will fit on one page, how many pages of the book will he fill?

7. $37 + 21 + 15 = ?$

8. Two figures with the same size and shape are called _________.

9. $7 \times 9 = ?$

10. Round *6,541* to the nearest thousand.

11. $5{,}216 + 3{,}827 = ?$

12. Which digit is in the hundreds place in the number *9,375*?

13. $355 \div 5 = ?$

14. List the even numbers that fall between 51 and 57.

15. $703 - 365 = ?$

16. If it is 6:25 now, what time will it be in 4 hours and 5 minutes?

17. Draw 2 similar rectangles.

18. $324 \times 4 = ?$

19. How many cups are in 4 pints?

20. Are these fractions equivalent?

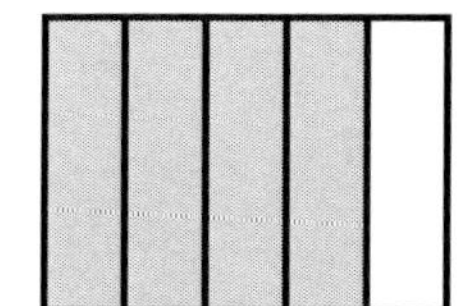

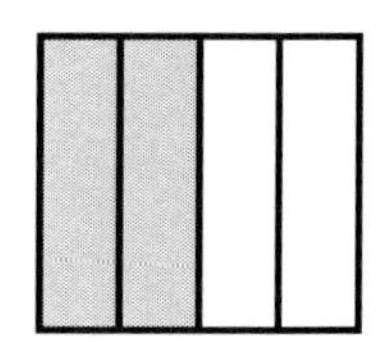

1.	2.	3.	4.
5.	6.	7.	8.
9.	10.	11.	12.
13.	14.	15.	16.
17.	18.	19.	20.

Lesson #3

1. Draw parallel lines.

2. Write the time shown on the clock.
3. $4 \times 6 = ?$
4. Write the fact family for 3, 6, and 18.
5. $4{,}854 - 1{,}229 = ?$

6. Draw a square and shade $\frac{3}{4}$ of it.
7. $47 \div 9 = ?$
8. $36 \times 3 = ?$
9. Draw a ray.
10. $2 \times 3 \times 5 = ?$
11. What will be the time 10 minutes before noon?
12. Shawna has 30 nickels. How much money does she have?
13. Write the name of this shape.

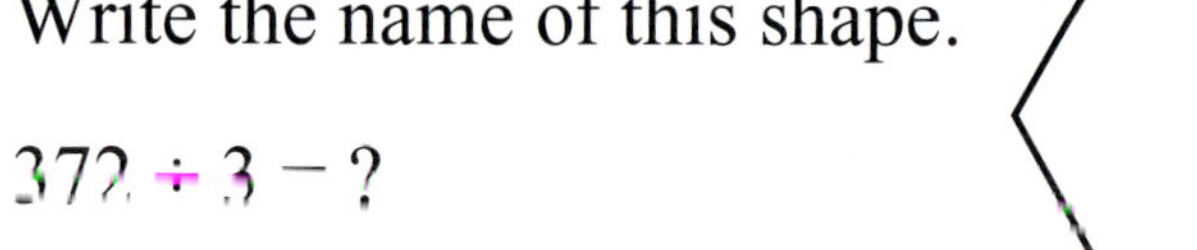

14. $372 \div 3 = ?$
15. Would a table weigh about 40 ounces or 40 pounds?
16. Round *85,166* to the nearest ten thousand.
17. How many months are in a year?
18. $356 + 198 = ?$
19. Which figure is at point (3, 1) on the graph?
20. Write the ordered pair for the ●.

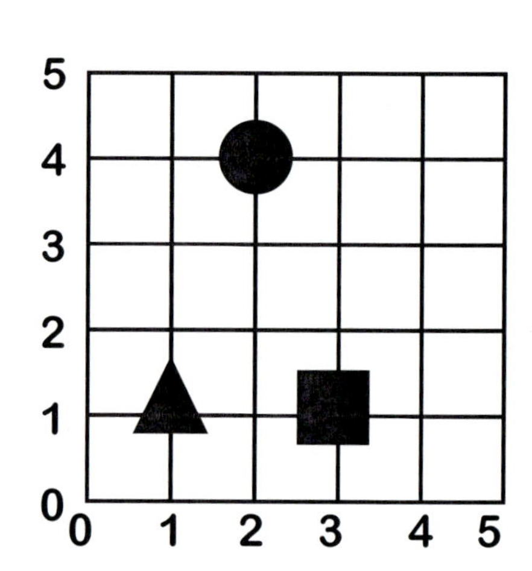

1.	2.	3.	4.
5.	6.	7.	8.
9.	10.	11.	12.
13.	14.	15.	16.
17.	18.	19.	20.

Lesson #4

1. 704 – 367 = ?

2. What month is 5 months after April?

3. 37 × 3 = ?

4. Draw a right angle.

5. 54 ÷ 4 = ?

6. 3,365 + 4,976 = ?

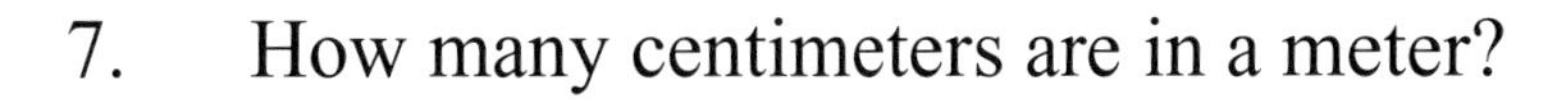

7. How many centimeters are in a meter?

8. Which digit is in the tens place in *7,064*?

9. 864 ◯ 846

10. Are these lines parallel?

11. 565 ÷ 5 = ?

12. What will be the time 30 minutes before 8:00?

13. Write the next number in the sequence. 35, 45, 55, …

14. Write *56,241* in expanded form.

15. 5 × 9 = ?

16. What fraction is shaded?

17. The answer to a multiplication problem is called the __________.

18. How many dimes are in 70¢?

19. The top number in a fraction is called the ___________.

20. There were 85 people on a bus. At the first stop, 18 people got off the bus. At the second stop, 21 people got off. How many people are still on the bus?

1.	2.	3.	4.
5.	6.	7.	8.
9.	10.	11.	12.
13.	14.	15.	16.
17.	18.	19.	20.

Lesson #5

1. $762 \div 2 = ?$

2. Write the standard number for *600 + 90 + 3*.

3. The sum is the answer to a(n) __________ problem.

4. Write the odd numbers between 30 and 38.

5. $8,321 + 4,769 = ?$

6. $2 \times 2 \times 4 = ?$

7. How much money is shown here?

8. $800 - 248 = ?$

9. Which digit is in the hundreds place in *8,352*?

10. Round *66,314* to the nearest thousand.

11. Which is longer, 3 yards or 3 miles?

12. $49 \div 7 = ?$

13. How many inches are in 2 feet?

14. $249 \times 2 = ?$

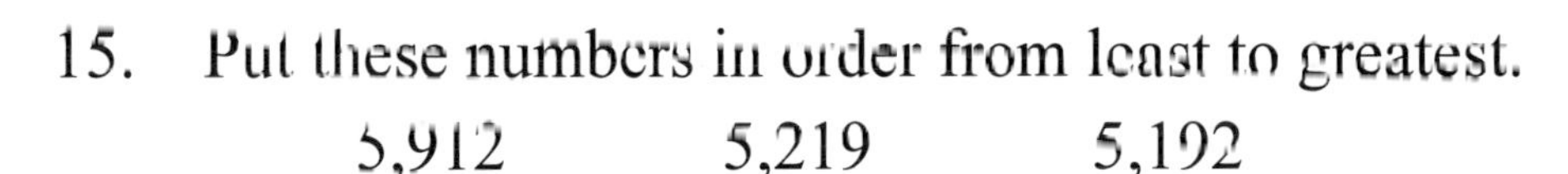

15. Put these numbers in order from least to greatest.
    5,912   5,219   5,192

16. Find the mode of the numbers 18, 25, 57, 25 and 16.

17. *A closed shape made up of line segments* is called a(n) ________.

18. Draw 2 congruent rectangles.

19. Kendall bought a video game for $55.95. If he gives the clerk $60.00, how much change will he get back?

20. How many hours are in a day?

1.	2.	3.	4.
5.	6.	7.	8.
9.	10.	11.	12.
13.	14.	15.	16.
17.	18.	19.	20.

Lesson #6

1. Write *2,543* using words.

2. Draw a pentagon.

3. 3,369 + 1,219 = ?

4. Which shaded fraction is greater?

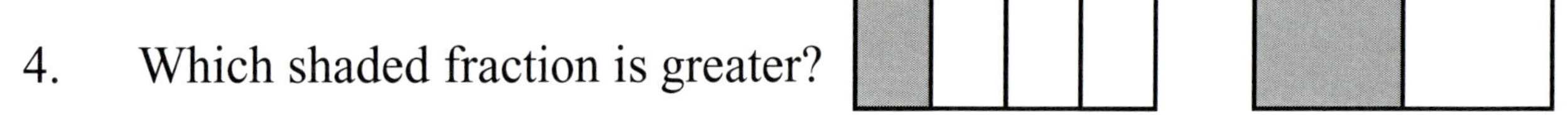

5. Round *3,298* to the nearest hundred.

6. 63,293 ◯ 63,329

7. The top number in a fraction is called the ____________.

8. How many days are in a year?

9. $525 \times 6 = ?$

10. Is the height of a door about 8 inches or 8 feet?

11. $372 \div 3 = ?$

12. Draw a square and shade $\frac{1}{4}$ of it.

13. Draw two congruent circles.

14. Find $\frac{1}{5}$ of 25. (Hint: How many groups of 5 are in 25?)

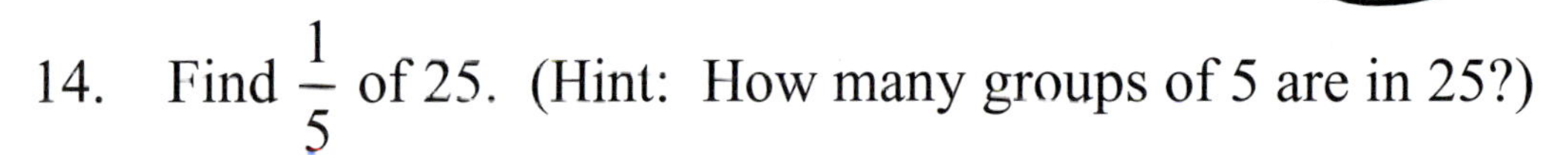

15. $21 \div 3 = ?$

16. Write a fact family for 4, 8 and 32.

17. $10 \times 10 = ?$

18. Ryan has 9 quarters. How much money does he have?

19. What time will it be in 15 minutes, if it is 2:45 now?

20. A small pizza has 475 calories. A hot dog has 245 calories. How many more calories are in the pizza than in the hot dog?

1.	2.	3.	4.
5.	6.	7.	8.
9.	10.	11.	12.
13.	14.	15.	16.
17.	18.	19.	20.

Lesson #7

1. Is the length of a dollar bill about 6 inches or 6 yards?

2. Round *72,416* to the nearest thousand.

3. $26 \div 3 = ?$

4. $6{,}320 - 3{,}765 = ?$

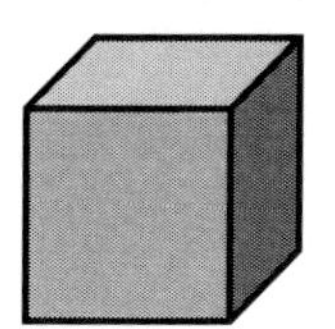

5. Name the shape.

6. What do we call the answer to a subtraction problem?

7. Write *60,000 + 4,000 + 900 + 70 + 3* in standard form.

8. $942 \div 2 = ?$

9. How many quarts are in 4 gallons?

10. $2 \times 3 \times 3 = ?$

11. Is this figure a polygon?

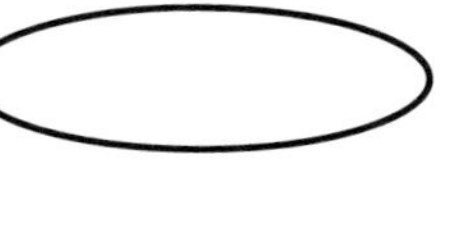

12. Find $\frac{1}{3}$ of 15.

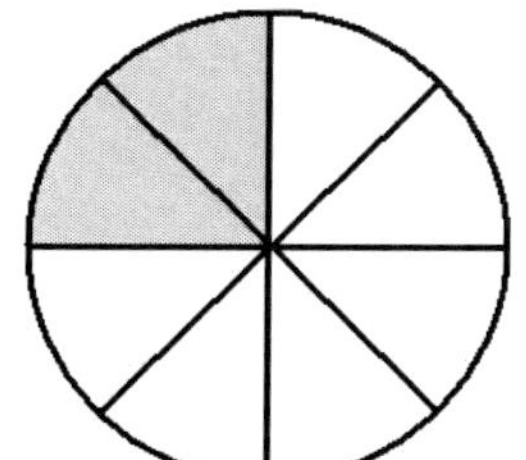

13. What fraction of the circle is shaded?

14. If it is 2:30 now, what time was it 4 hours ago?

15. $\frac{5}{9} - \frac{1}{9} = ?$

16. Over the summer Angie read 7 more books than Chris. Chris read 5 more books than Valerie did. Valerie read 10 books. How many books did Angie read?

17. $94 \times 3 = ?$

18. Is *257* an even number or an odd number?

19. Draw a right angle.

20. $246 + 445 = ?$

1.	2.	3.	4.
5.	6.	7.	8.
9.	10.	11.	12.
13.	14.	15.	16.
17.	18.	19.	20.

Lesson #8

1. Draw a square and show a line of symmetry on it.
2. $288 \div 3 = ?$
3. Would a swimming pool be about 60 feet deep or 6 feet deep?
4. Write the time.

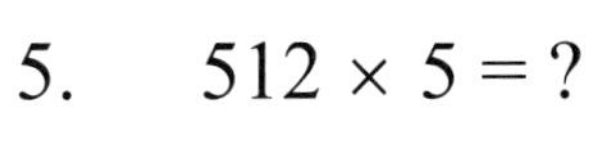

5. $512 \times 5 = ?$

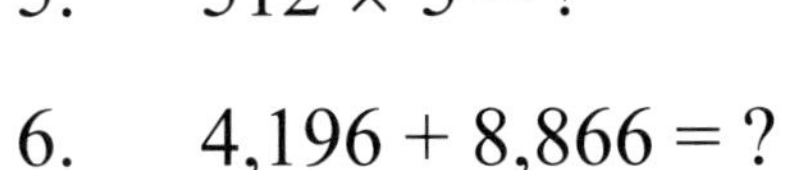

6. $4{,}196 + 8{,}866 = ?$
7. Which digit is in the ten thousands place in *654,713*?
8. How many feet are in a yard?
9. Write $\frac{3}{10}$ using words.
10. $6 \times 8 = ?$
11. Are these lines parallel?
12. $\$57.29 - \$17.56 = ?$

13. $400 - 69 = ?$

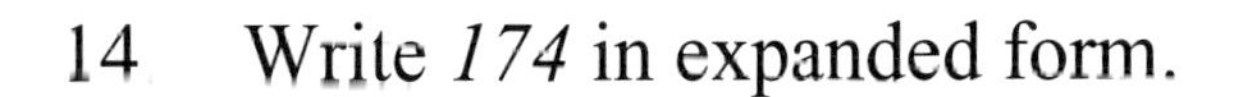

14. Write *174* in expanded form.
15. How many cups are in a pint?
16. A hexagon has ________ sides.
17. Is *2,363* an even or an odd number?
18. Draw a ray.
19. How many monkeys are at the zoo?
20. How many more bears than elephants are at the zoo?

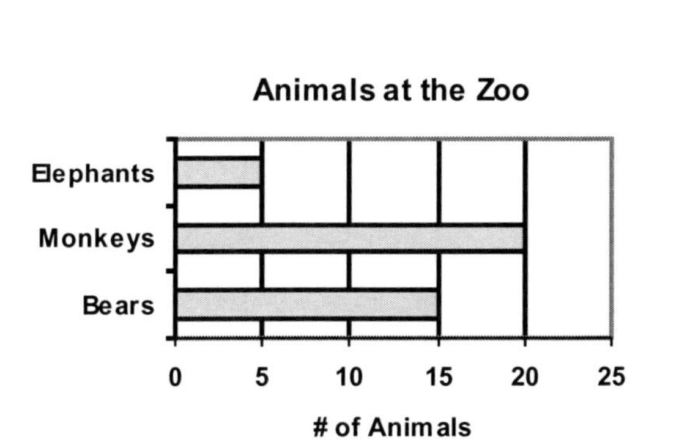

1.	2.	3.	4.
5.	6.	7.	8.
9.	10.	11.	12.
13.	14.	15.	16.
17.	18.	19.	20.

Lesson #9

1. Figures with the same size and shape are called _________.
2. 872 ÷ 4 = ?
3. Round *67* to the nearest ten.
4. Draw a line segment.
5. How many hours are in 3 days?
6. 35 + 16 + 29 = ?
7. Which digit is in the thousands place in *37,215*?

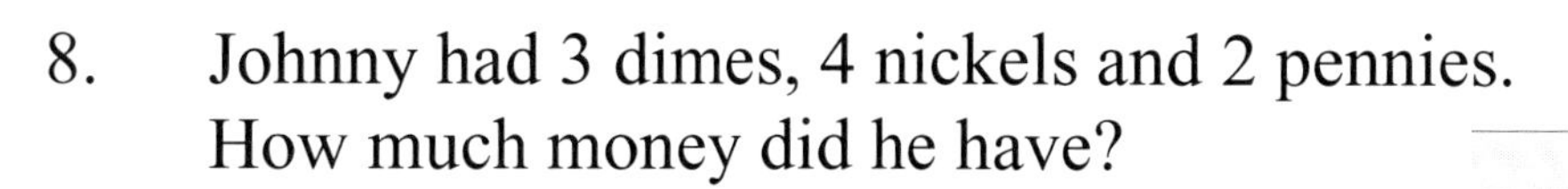

8. Johnny had 3 dimes, 4 nickels and 2 pennies.
   How much money did he have?

9. 8 × 9 = ?

10. A five-sided polygon is called a _________.
11. 436 × 6 = ?
12. What fraction is shaded?

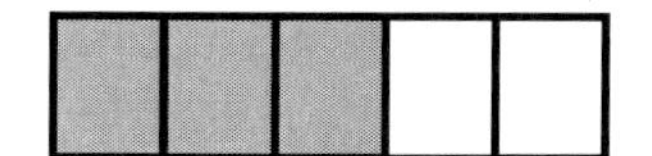

13. 93 ÷ 3 = ?
14. Write the next number in the sequence. 12, 20, 28, ...
15. Look at the shape to the right. Write its name.

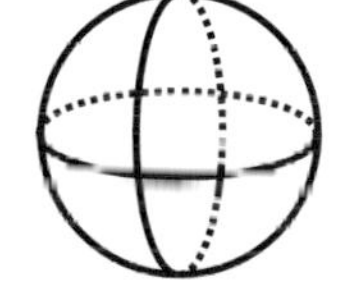

16. It is 5:10 now. What time was it 6 hours ago?
17. $\frac{1}{7} + \frac{3}{7} = ?$
18. The bottom number in a fraction is called the ___________.
19. Harold collects football cards. He gave 23 cards to his friend and 10 cards to his brother. He has 75 cards left. How many cards did Harold have before he gave some away?
20. 308 – 167 = ?

1.	2.	3.	4.
5.	6.	7.	8.
9.	10.	11.	12.
13.	14.	15.	16.
17.	18.	19.	20.

Lesson #10

1. How many centimeters are in a meter?

2. $94 \times 7 = ?$

3. An eight-sided polygon is called a(n) _________.

4. Find the mode of 77, 39, 22, 19 and 39.

5. What fraction of the circle is shaded?

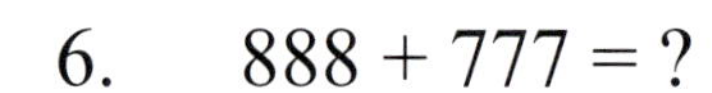

6. $888 + 777 = ?$

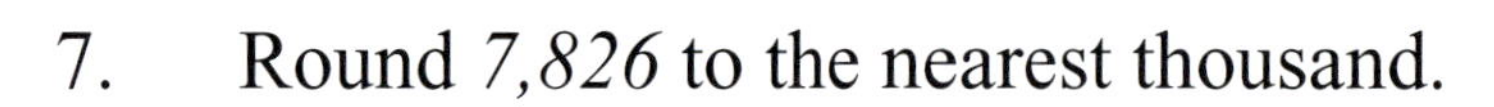

7. Round *7,826* to the nearest thousand.

8. How many years are in a decade?

9. Draw two congruent squares.

10. Which digit is in the ten thousands place in *26,514*?

11. $900 - 392 = ?$

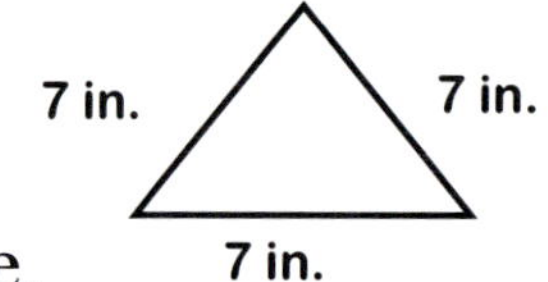

12. Find the perimeter of this triangle.

13. Write the standard number for $800 + 40 + 2$.

14. Find $\frac{1}{5}$ of 40.

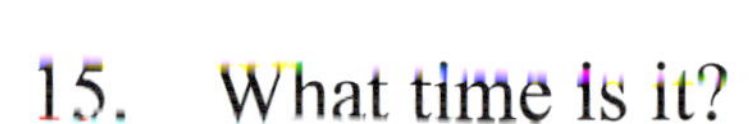

15. What time is it?

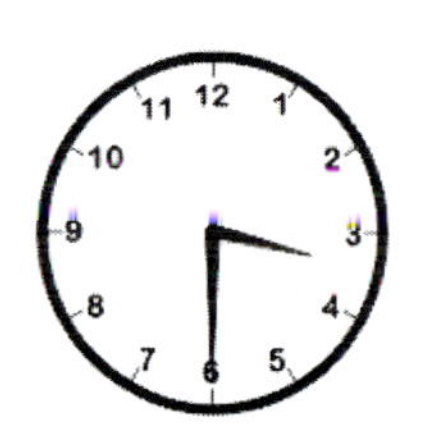

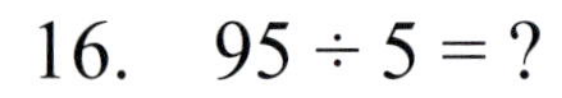

16. $95 \div 5 = ?$

17. What is the name of the shape?

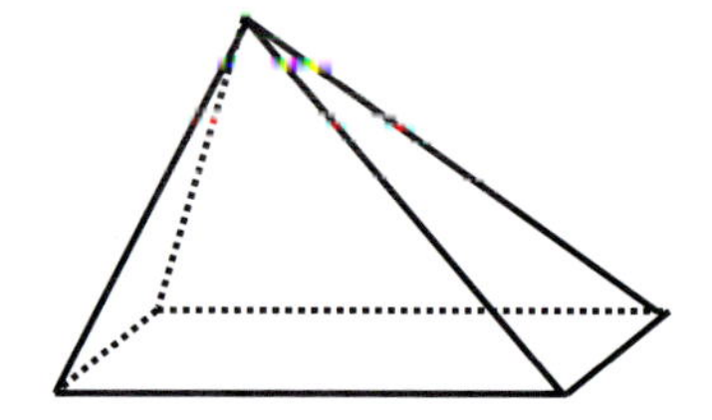

18. Mr. Thomas has two pieces of wire. One piece of wire is 9 feet long. The length of both pieces of wire together is 20 feet. How long is the other piece of wire?

19. The answer to an addition problem is called the ___________.

20. There are ______________ feet in a yard.

1.	2.	3.	4.
5.	6.	7.	8.
9.	10.	11.	12.
13.	14.	15.	16.
17.	18.	19.	20.

Lesson #11

1. Which digit is in the hundreds place in *36,507*?
2. 355 + 992 = ?
3. What time is it on this clock?

4. 9 × 8 = ?
5. The answer to a division problem is called the __________.
6. 8,366 + 4,597 = ?
7. Name the shape.
8. Draw 2 congruent circles.
9. 512 – 264 = ?
10. What will be the time 10 minutes before 4:00?
11. Round *736* to the nearest ten.
12. 7,000 – 2,433 = ?
13. 36 ÷ 4 = ?
14. How many seconds are in 4 minutes?
15. 72 × 8 = ?
16. Is a tree more likely to be 18 feet tall or 18 miles tall?
17. 16 – 7 = ?
18. How many quarters are in $3?
19. 38 ÷ 5 = ?
20. How many chose Reading as their favorite pastime?

    How many chose watching TV and playing video games as their favorites?

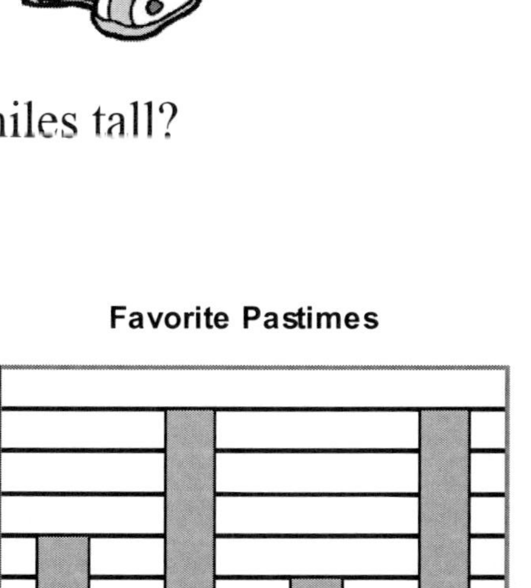

1.	2.	3.	4.
5.	6.	7.	8.
9.	10.	11.	12.
13.	14.	15.	16.
17.	18.	19.	20.

Lesson #12

1. 3,816 + 5,774 = ?

2. Round *1,568* to the nearest thousand.

3. How many feet are in a mile?

4. 60,000 + 5,000 + _______ + 50 + 2 = 65,352

5. Name the figure.

6. Draw intersecting lines.

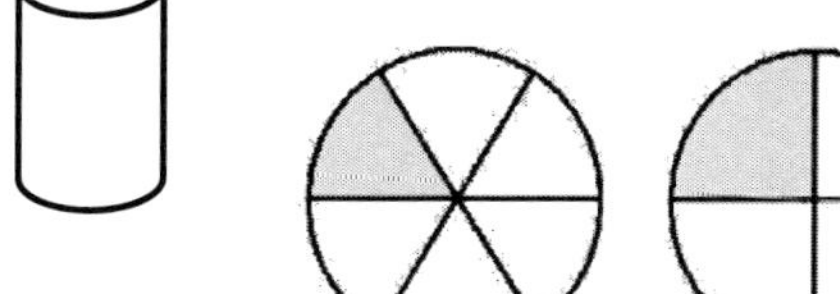

7. What shaded part is larger?

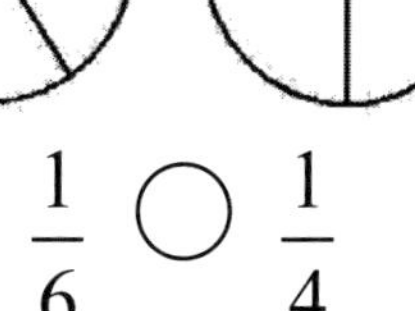

$\frac{1}{6} \bigcirc \frac{1}{4}$

8. 9,364 – 5,175 = ?

9. The product of two factors is 42. One factor is 6. What is the other factor?

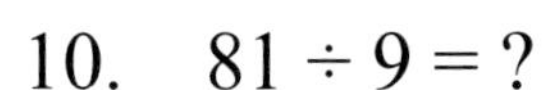

10. 81 ÷ 9 = ?

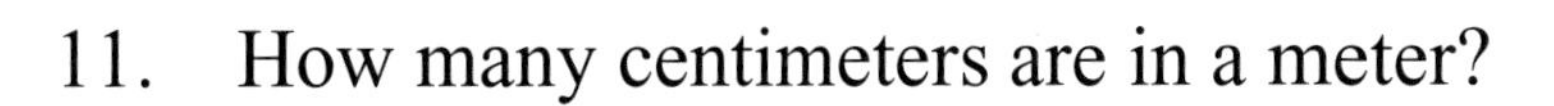

11. How many centimeters are in a meter?

12. 5 × 7 = ?

13. Does a leaf weigh about 1 ounce or 1 pound?

14. The top number in a fraction is called the ___________.

15. What time will it be in 6 hours if it is 7:15 now?

16. 456 ÷ 4 = ?

17. Draw 2 similar circles.

18. How much money is 5 quarters?

19. On which day of the week is November 16th?

20. What is the date two weeks after November 6th?

November

S	M	T	W	T	F	S
				1	2	3
4	5	6	7	8	9	10
11	12	13	14	15	16	17
18	19	20	21	22	23	24
25	26	27	28	29	30	

1.	2.	3.	4.
5.	6.	7.	8.
9.	10.	11.	12.
13.	14.	15.	16.
17.	18.	19.	20.

Lesson #13

1. A closed figure made up of line segments is called ___________.

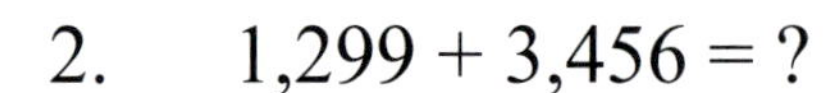

2. 1,299 + 3,456 = ?

3. Draw a line segment.

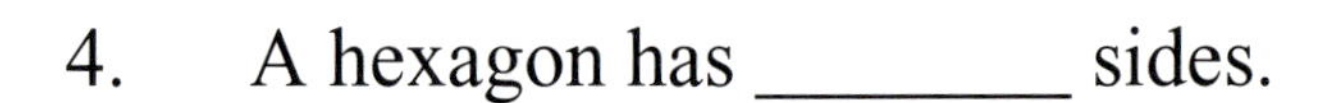

4. A hexagon has ________ sides.

5. Write the time shown on the clock.

6. Round *832* to the nearest hundred.

7. $37 \times 7 = ?$

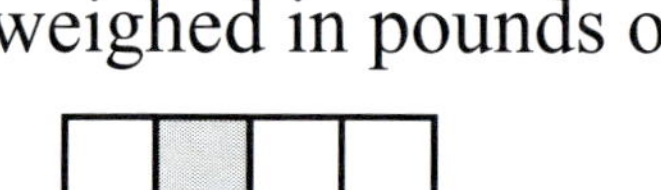

8. Would an elephant best be weighed in pounds or in tons?

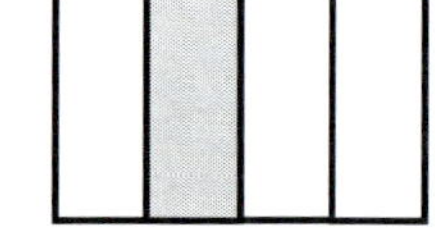

9. What fraction is shaded?

10. $6 \times 9 = ?$

11. Write the first five odd numbers.

12. Write the standard number for *7,000 + 500 + 30 + 7*.

13. $635 \div 5 = ?$

14. How many minutes are in 5 hours?

15. $24 \div 6 = ?$

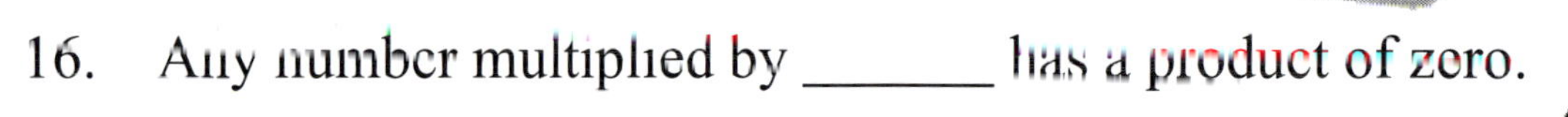

16. Any number multiplied by _______ has a product of zero.

17. How many cups are in 3 pints?

18. Put these numbers in order from greatest to least.

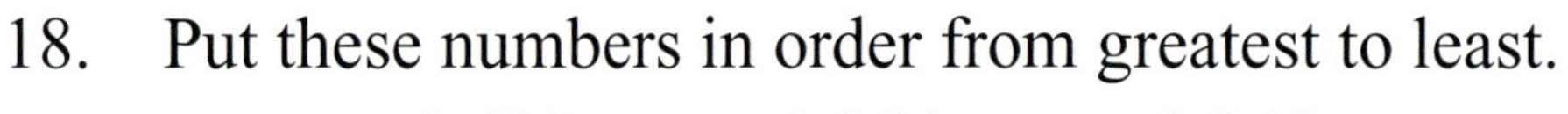

9,685 9,856 9,865

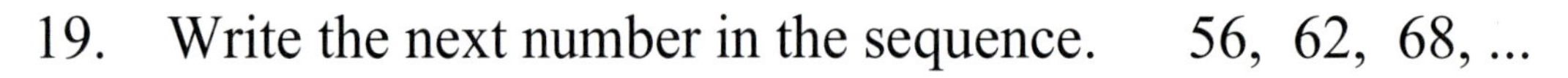

19. Write the next number in the sequence. 56, 62, 68, ...

20. Mrs. Garner used 3 feet of ribbon to decorate one package, 2 feet for another and 5 feet for a third package. She had 12 feet of ribbon left. How many feet of ribbon did she begin with?

1.	2.	3.	4.
5.	6.	7.	8.
9.	10.	11.	12.
13.	14.	15.	16.
17.	18.	19.	20.

Lesson #14

1. Write a fact family for 4, 5, and 20.
2. $785 + 442 = ?$
3. Round *364,299* to the nearest hundred thousand.
4. Find $\frac{1}{2}$ of 16.

5. $932 \times 4 = ?$
6. Write *32,471* in expanded form.
7. What will be the time 5 minutes before noon?

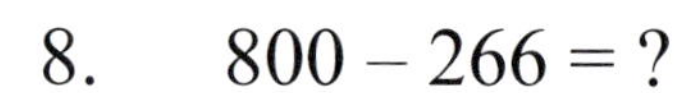

8. $800 - 266 = ?$

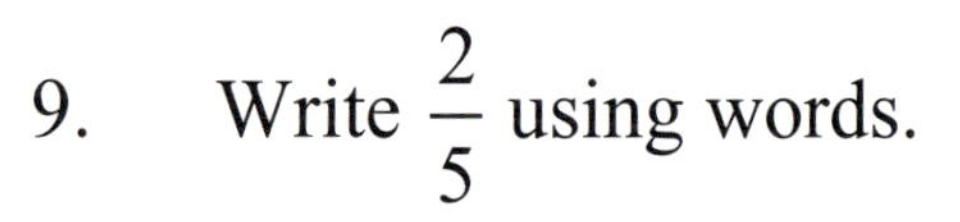

9. Write $\frac{2}{5}$ using words.

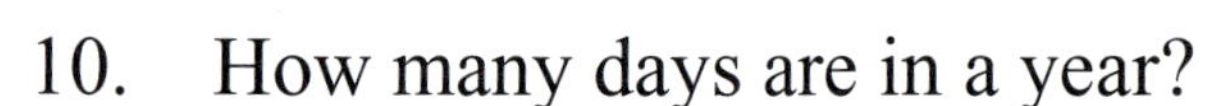

10. How many days are in a year?

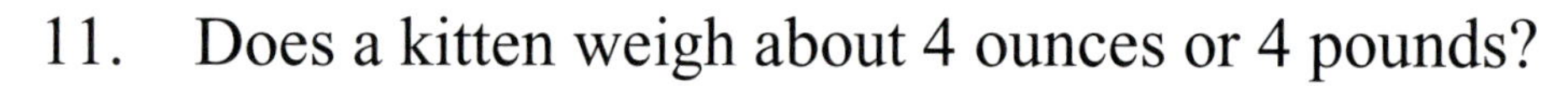

11. Does a kitten weigh about 4 ounces or 4 pounds?
12. $8 \times 8 = ?$

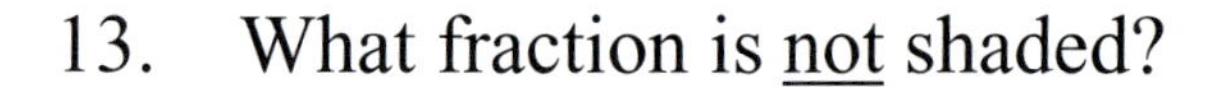

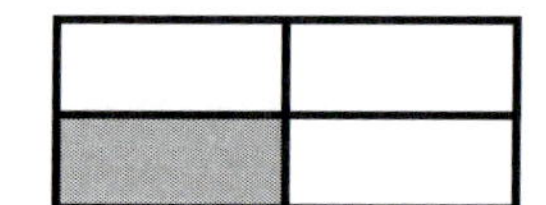

13. What fraction is <u>not</u> shaded?
14. 8,364 ◯ 8,634
15. $96 \div 3 = ?$
16. How many quarts are in 2 gallons?
17. Show a line of symmetry on a heart.

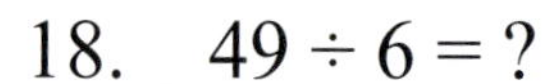

18. $49 \div 6 = ?$

19. Two figures with the same shape, but different sizes are ________.
20. Marissa is taking her family to see a play. There are two adults and three children in her family. The cost of an adult ticket is \$8 and the cost of a child's ticket is \$4. How much money will Marissa spend on tickets?

1.	2.	3.	4.
5.	6.	7.	8.
9.	10.	11.	12.
13.	14.	15.	16.
17.	18.	19.	20.

Lesson #15

1. $852 \div 2 = ?$

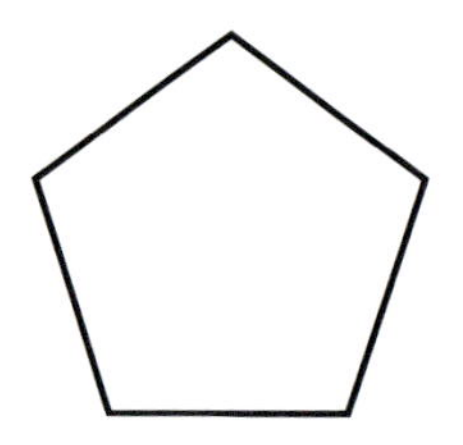

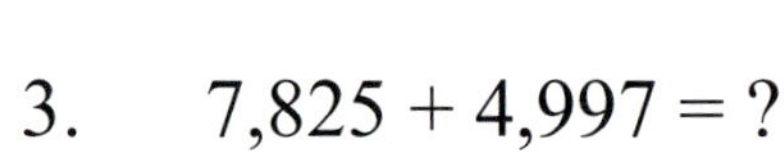

2. Write the name of the shape.

3. $7{,}825 + 4{,}997 = ?$

4. The product is the answer to a(n) __________ problem.

5. $2 \times 5 \times 3 = ?$

6. Stephanie has 3 quarters, 2 dimes and 3 nickels. How much money does she have?

7. Round *46,719* to the nearest ten thousand.

8. How many feet are in a yard?

9. Find the mode of this set of numbers. 21, 52, 13, 80, 52

10. Which is longer, 3 yards or 3 miles?

11. $2{,}538 - 963 = ?$

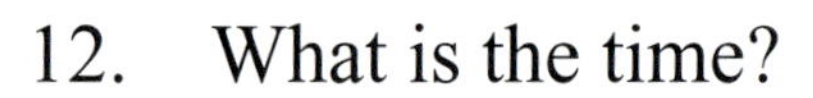

12. What is the time?

13. $8 \times 8 = ?$

14. How many cups are in 4 pints?

15. In a multiplication problem, the two numbers that are multiplied together are called ___________.

16. $216 \times 3 = ?$

17. Is this figure a polygon?

18. $27 \div 3 = ?$

19. Write the missing numbers in the sequence. 27, 29, ____, 33, ____

20. Kelsey made 3 dozen cupcakes for her birthday. There are 30 students in her class. If every student received a cupcake, how many cupcakes were left over?

1.	2.	3.	4.
5.	6.	7.	8.
9.	10.	11.	12.
13.	14.	15.	16.
17.	18.	19.	20.

Lesson #16

1. Write *4,353* using words.

2. What fraction is shaded?
3. $6 \times 6 = ?$
4. $4{,}207 - 2{,}973 = ?$

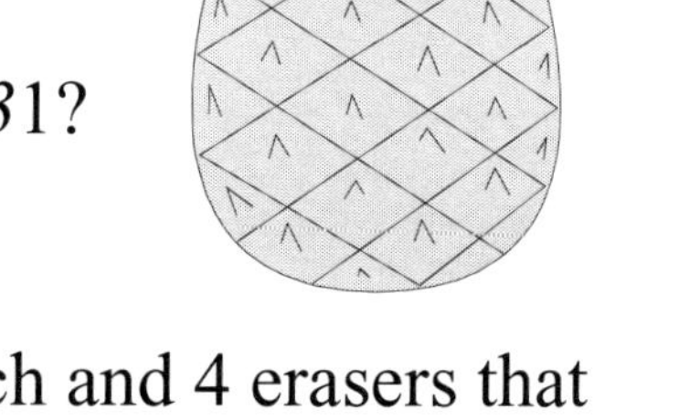

5. $524 \times 2 = ?$
6. How many days are in 7 weeks?
7. 62,907 ◯ 62,790
8. $972 \div 3 = ?$
9. Which digit is in the thousands place in *86,931*?
10. Draw a right angle.
11. Christina bought 5 pencils that cost \$0.08 each and 4 erasers that cost \$0.04 each. How much money did she spend on her supplies?
12. Write *three-fourths* as a fraction.
13. The sum is the answer to a(n) ___________ problem.
14. If it is 2:45 now, what time will it be in 7 hours and 5 minutes?
15. $5{,}036 + 8{,}879 = ?$
16. Find $\frac{1}{4}$ of 16.
17. $56 \div 8 = ?$
18. Does a pineapple weigh about 3 pounds or 3 ounces?
19. Write the name of the shape.

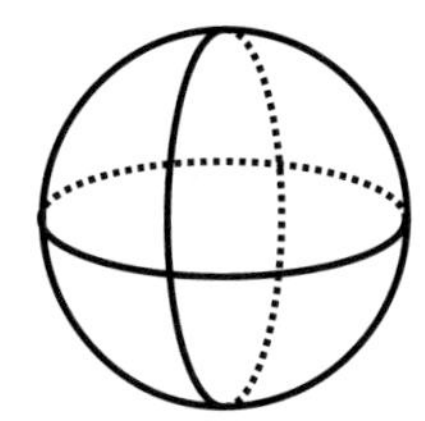

20. $\frac{5}{8} - \frac{2}{8} = ?$

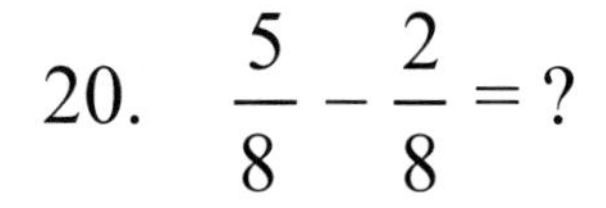

1.	2.	3.	4.
5.	6.	7.	8.
9.	10.	11.	12.
13.	14.	15.	16.
17.	18.	19.	20.

Lesson #17

1. Write *30,000 + 7,000 + 500 + 20 + 8* in standard form.
2. $26 \div 4 = ?$
3. $3 \times 7 = ?$
4. Is *4,216* an even number or an odd number?
5. Round *86,207* to the nearest ten thousand.
6. How many inches are in 3 feet?
7. $65.99 – $27.37 = ?
8. If it is 12:30 now, what time was it 4 hours ago?
9. $72 \times 3 = ?$
10. $6{,}185 + 7{,}236 = ?$
11. $424 \div 3 = ?$
12. $\frac{6}{7} - \frac{2}{7} = ?$
13. How many feet are in a mile?
14. Which digit is in the ten thousands place in *169,347*?
15. Draw 2 congruent squares.
16. Find the mode of 51, 27, 34, 27 and 60.
17. Is this an obtuse angle?

18. $600 \times 9 = ?$
19. Jerome shot the basketball 10 times. He made 7 baskets. In how many of his shots did he make a basket? Write your answer as a fraction.
20. List the even numbers between 41 and 49.

1.	2.	3.	4.
5.	6.	7.	8.
9.	10.	11.	12.
13.	14.	15.	16.
17.	18.	19.	20.

Lesson #18

1. $\frac{1}{5} + \frac{1}{5} = ?$

2. 418 – 89 = ?

3. An octagon has ___________ sides.

4. Chris has 2 dollars, 2 quarters and 3 dimes. How much money does he have?

5. Round *456* to the nearest ten.

6. 34 ÷ 4 = ?

7. How many feet are in a yard?

8. A closed figure made up of line segments is a(n) ________.

9. 36,517 ◯ 36,175

10. The answer to a division problem is the ___________.

11. Write the first four even numbers.

12. Which digit is in the hundred thousands place in *451,368*?

13. Write the standard number for *900,000 + 40,000 + 6,000 + 500 + 20 + 7*.

14. 15 – 6 = ?

15. Write the time shown on the clock.

16. Which month comes 6 months after December?

17. Write the missing numbers. 14, 20, ____, 32, ____

18. In the fraction $\frac{2}{3}$, which number is the denominator?

19. Lauren has twice as many blocks as Sean. If Sean has 17 blocks, how many blocks does Lauren have?

20. 8,553 + 9,871 = ?

1.	2.	3.	4.
5.	6.	7.	8.
9.	10.	11.	12.
13.	14.	15.	16.
17.	18.	19.	20.

Lesson #19

1. What will be the time 30 minutes after 7:00?
2. Draw 2 similar rectangles.
3. $7{,}716 + 8{,}947 = ?$
4. $872 \div 4 = ?$
5. Draw a line segment.
6. Round *$9.74* to the nearest dollar.
7. $800 - 256 = ?$
8. Which digit is in the thousands place in *31,246*?

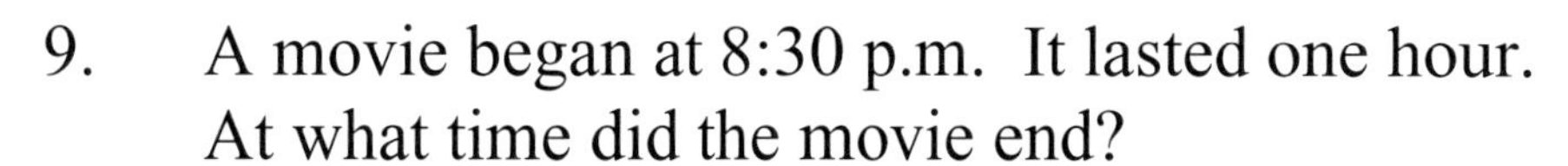

9. A movie began at 8:30 p.m. It lasted one hour.
   At what time did the movie end?
10. Keith has 9 dimes, 1 nickel and 3 pennies.
    How much money does Keith have?
11. Find $\frac{1}{3}$ of 15.
12. $536 \times 6 = ?$
13. Would the length of a car be about 7 inches or 7 feet?
14. $(2 \times 3) \times 5 = ?$
15. $72 \div 9 = ?$
16. Write the odd numbers between 84 and 90.
17. A polygon with six sides is called a(n) __________.
18. $8 \times 6 = ?$
19. Susan has 36 books. If she has 4 shelves and she puts the same number of books on each shelf, how many books will be on each shelf?
20. What fraction is shaded?

1.	2.	3.	4.
5.	6.	7.	8.
9.	10.	11.	12.
13.	14.	15.	16.
17.	18.	19.	20.

Lesson #20

1. $98 \times 7 = ?$

2. Round *832,175* to the nearest hundred thousand.

3. $373 + 927 = ?$

4. How many cups are in 4 pints?

5. $5 \times 6 = ?$

6. Are these shapes congruent?

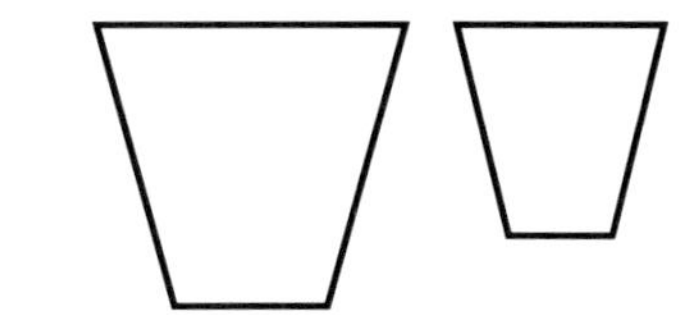

7. Ricardo had 2 quarters, 4 dimes and 1 nickel. How much money did he have?

8. $646 \div 2 = ?$

9. Write the time.

10. The difference is the answer to a(n) _________ problem.

11. Write a fact family for 5, 7 and 35.

12. The numerator is the __________ number in a fraction.

13. $8{,}000 - 2{,}974 = ?$

14. Jason is 3 years older than Julie. Julie is 2 years younger than Todd. Todd is 11 years old. How old is Jason?

15. Would a grape more likely weigh about 1 ounce or 1 pound?

16. Draw a ray.

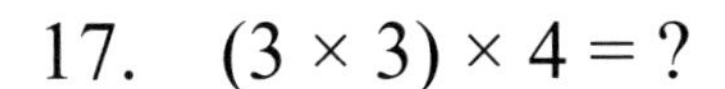

17. $(3 \times 3) \times 4 = ?$

18. $54 \div 6 = ?$

19. What animal is located at (4, 2)?

20. The lion is located at what point?

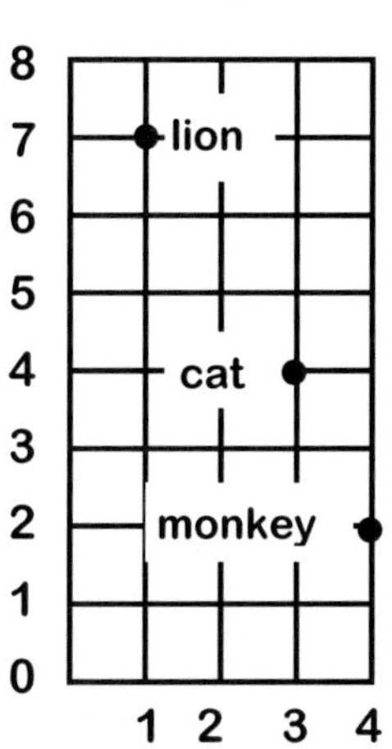

1.	2.	3.	4.
5.	6.	7.	8.
9.	10.	11.	12.
13.	14.	15.	16.
17.	18.	19.	20.

Lesson #21

1. Which is greater, 3 quarters or 5 dimes?

2. Carina went for a walk at 4:30. She got home at 5:10. How long was her walk?

3. $\frac{5}{8} - \frac{2}{8} = ?$

4. Round *347,251* to the nearest ten thousand.

5. Any number multiplied by zero has a product of _________.

6. Draw 2 similar pentagons.

7. 8,893 + 7,565 = ?

8. 126 ÷ 3 = ?

9. What will be the time 2 hours before noon?

10. Write the missing numbers in the sequence. 8, 12, ____, 20, ____

11. (2 × 4) × 5 = ?

12. A four-sided polygon is called a(n) _________________.

13. Round *4,851* to the nearest thousand.

14. 5,831 ◯ 5,381

15. What fraction of the rectangle is shaded?

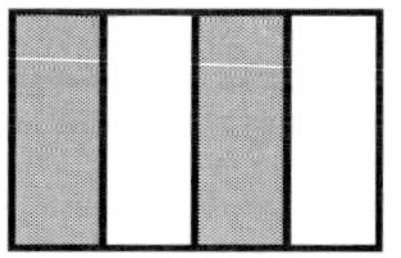

16. 42 ÷ 6 = ?

17. Name the shape.

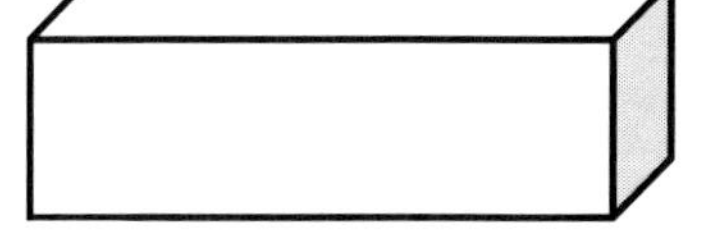

18. 44 × 5 = ?

19. 6 + 7 = ?

20. Brandon bought a pen for \$0.35 and a notebook for \$1.45. He gave the clerk \$2.00. How much change did Brandon get back?

1.	2.	3.	4.
5.	6.	7.	8.
9.	10.	11.	12.
13.	14.	15.	16.
17.	18.	19.	20.

Lesson #22

1. Draw 2 rhombuses (diamonds) that are congruent.
2. $41 + 15 + 27 = ?$
3. If it is 2:20 now, what time was it 5 hours ago?
4. Write *seven thousand, six hundred eighty-two* as a standard number.
5. $7,206 - 2,541 = ?$
6. List the odd numbers between 50 and 58.
7. Find the mode of 12, 31, 16, 47, 12, and 53.
8. Round *53,937* to the nearest ten.
9. $231 \times 4 = ?$
10. The denominator is the __________ number in a fraction.
11. How many quarters are in $7?
12. Sixty-three children chose red as their favorite color. Forty-seven children chose blue. How many more children chose red than blue?
13. The product of two factors is 63. One factor is 9. What is the other factor?
14. $4 \times 7 = ?$
15. How many quarts are in 6 gallons?
16. $68 \div 7 = ?$
17. Which digit is in the ten thousands place in *139,274*?
18. How many centimeters are in a meter?
19. Would a stove best be weighed in ounces or in pounds?
20. There were 10 problems on the science quiz. Sasha got 3 wrong. What fraction of the problems did she get wrong? What fraction did she get correct?

1.	2.	3.	4.
5.	6.	7.	8.
9.	10.	11.	12.
13.	14.	15.	16.
17.	18.	19.	20.

Lesson #23

1. $(15 \div 3) + 4 = ?$

2. $5 \times 6 = ?$

3. $3{,}126 - 887 = ?$

4. Write a fact family for 8, 9 and 72.

5. $935 \times 5 = ?$

6. How many ounces are in 2 pounds?

7. Round *9,317* to the nearest thousand.

8. Find $\frac{1}{6}$ of 24.

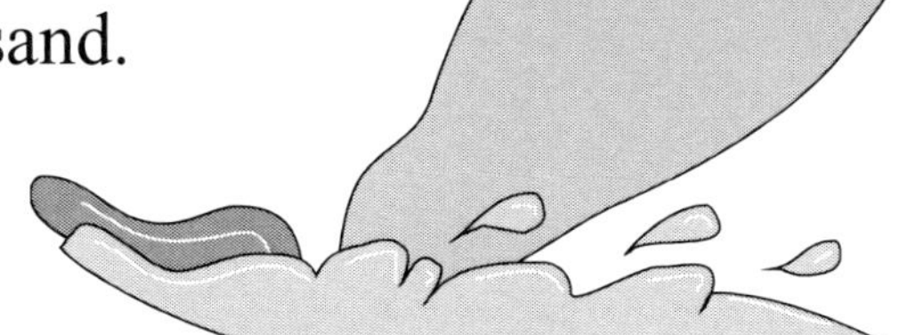

9. Write the decimal *0.45* using words.

10. Write *35,687* in expanded form.

11. $\frac{7}{8} - \frac{2}{8} = ?$

12. What is the name of the shape?

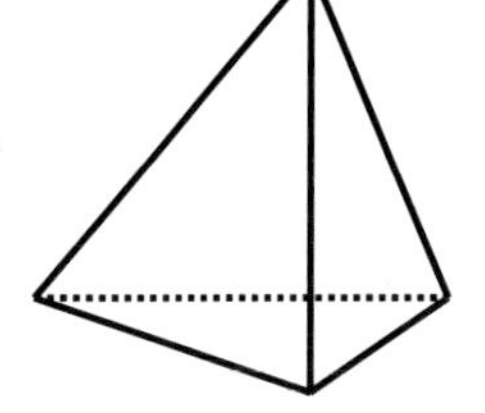

13. $8 \times 8 = ?$

14. What will be the time 1 hour and 30 minutes after 7:00?

15. $735 \div 5 = ?$

16. The quotient is the answer to a(n) __________ problem.

17. $667 + 325 = ?$

18. Nine dimes and 3 nickels is how much money?

19. Put these numbers in order from least to greatest.

    5,319 3,976 9,212

20. How many years are in 3 decades?

1.	2.	3.	4.
5.	6.	7.	8.
9.	10.	11.	12.
13.	14.	15.	16.
17.	18.	19.	20.

Lesson #24

1. 8,305 + 6,298 = ?

2. Study the table. How much will 5 gallons of gasoline cost? What is the cost of 8 gallons?

Cost of Gasoline

Gallons	1	2	3	4
Price	$3	$6	$9	$12

3. 3 × 2 × 4 = ?

4. 83 ÷ 9 = ?

5. 322 ÷ 2 = ?

6. The *number that comes up most often in a set of numbers* is called the ___________.

7. The product of two numbers is 42. One of the numbers is 7. What is the other number?

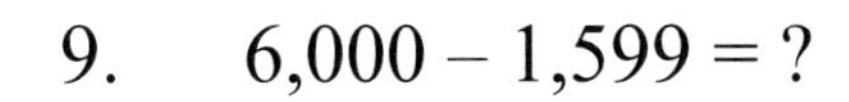

8. (12 ÷ 4) + 6 = ?

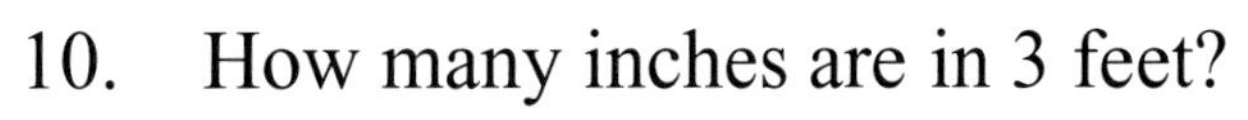

9. 6,000 – 1,599 = ?

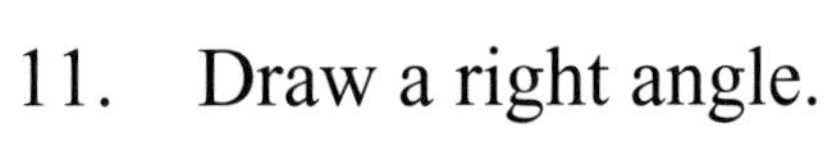

10. How many inches are in 3 feet?

11. Draw a right angle.

12. Round *186,255* to the nearest hundred thousand.

13. Which digit is in the hundreds place in *362,194*?

14. Jeffery bought 2 cans of paint for $7 each and a brush for $5.75. How much did he spend? How much change would he get back from a $20 bill?

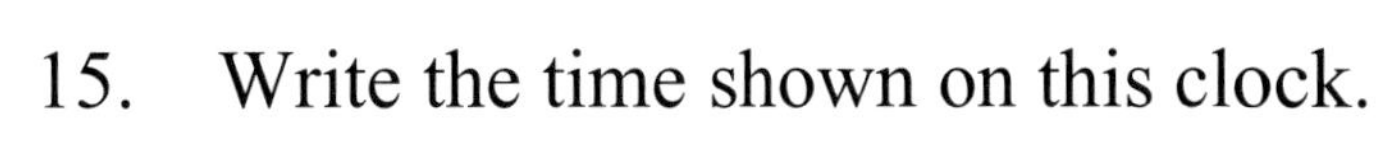

15. Write the time shown on this clock.

16. 4,076 × 6 = ?

17. Marsha has 3 coins that add up to 45¢. What are the coins?

18. 6 + 9 = ?

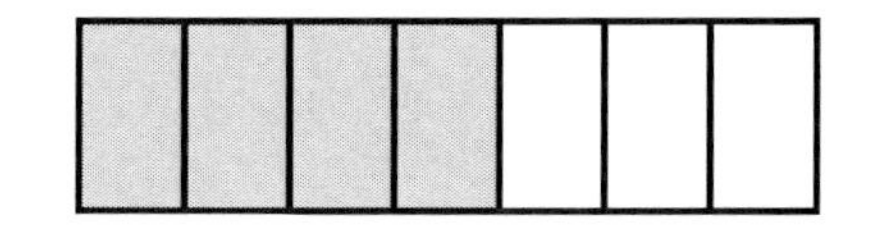

19. What fraction of the rectangle is shaded?

20. A pie was divided into 6 equal pieces. What fraction of the pie was left if Ty ate $\frac{1}{6}$ of the pie and Brad ate $\frac{3}{6}$ of the pie?

1.	2.	3.	4.
5.	6.	7.	8.
9.	10.	11.	12.
13.	14.	15.	16.
17.	18.	19.	20.

Lesson #25

1. $16 \times 6 = ?$

2. Write the even numbers between 61 and 69.

3. $2{,}354 + 7{,}917 = ?$

4. Would a skunk weigh about 16 ounces or 16 pounds?

5. What fraction is shaded?

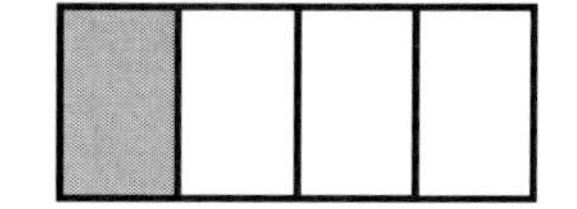

6. $(3 \times 7) \times 3 = ?$

7. Draw intersecting lines.

8. Antonio left for school at 8:15. He arrived at school at 8:45. How long did it take him to get to school?

9. $700 - 213 = ?$

10. Find the perimeter of the figure to the right.

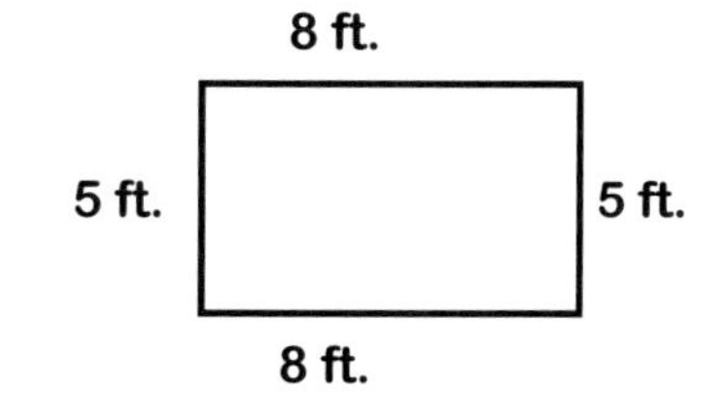

11. How many feet are in a mile?

12. $96 \div 4 = ?$

13. How many years are in 4 centuries?

14. Round *32,186* to the nearest ten thousand.

15. There are 32 students in Mrs. Lender's class. If Mrs. Lender put her students into 4 equal groups for a project, how many students would be in each group?

16. Name the shape.

17. $4 \times 8 = ?$

18. Write the next number in the sequence. 19, 23, 27, _____

19. Which digit is in the thousands place in *35,690*?

20. $\frac{2}{5} + \frac{2}{5} = ?$

1.	2.	3.	4.
5.	6.	7.	8.
9.	10.	11.	12.
13.	14.	15.	16.
17.	18.	19.	20.

Lesson #26

1. $6{,}000 \times 5 = ?$

2. How many months are in a year?

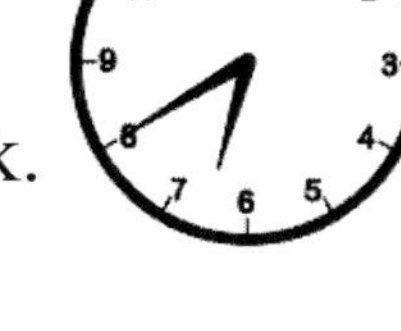

3. Write the time shown on this clock.

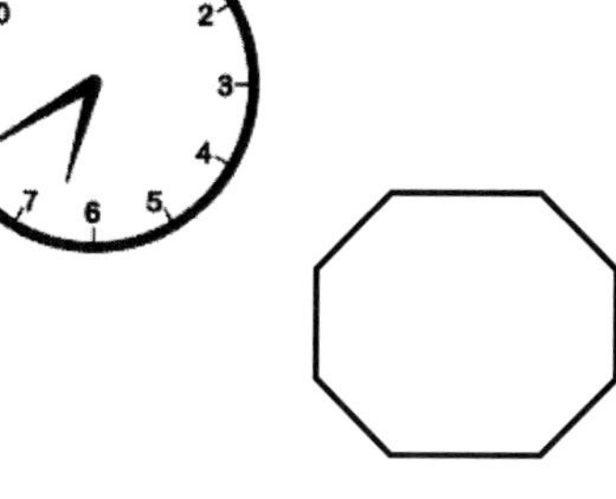

4. Write the name of the polygon.

5. Is *358* an even or an odd number?

6. Round *93* to the nearest ten.

7. $188 \div 2 = ?$

8. If it is 9:10, what time was it 5 hours and 10 minutes ago?

9. $3{,}317 + 4{,}588 = ?$

10. Write *26,489* in expanded form.

11. How much money is 6 quarters?

12. $605 - 371 = ?$

13. Which digit is in the thousands place in *718,254*?

14. Draw 2 congruent hearts.

15. 65,386 ◯ 56,863

16. Name the denominator in the fraction, $\frac{5}{8}$.

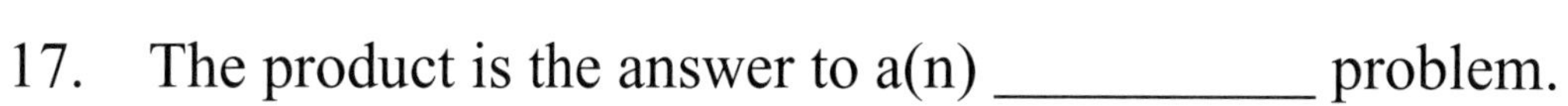

17. The product is the answer to a(n) __________ problem.

18. Paula read 28 pages of her library book the first day, 16 pages the second day, and 34 pages the third day. She still has 64 pages to read. How many pages are in her library book?

19. Which is the better temperature for going swimming, 59°F or 85°F?

20. What is the mode of 34, 56, 78, 56 and 93?

1.	2.	3.	4.
5.	6.	7.	8.
9.	10.	11.	12.
13.	14.	15.	16.
17.	18.	19.	20.

Lesson #27

1. Which digit is in the ten thousands place in *168,032*?

2. A flight from Cleveland, Ohio to Orlando, Florida should take about 2 hours or 2 days?

3. $99 \div 3 = ?$

4. Round *\$6.54* to the nearest dollar.

5. $63 \div 7 = ?$

6. The answer to an addition problem is the __________.

7. What will be the time 30 minutes after 12:15?

8. $530 \times 5 = ?$

9. $2{,}671 + 4{,}765 = ?$

10. Write the odd numbers between 80 and 90.

11. How many inches are in 4 feet?

12. Give *300,000 + 50,000 + 4,000 + 800 + 90 +3* as a standard number.

13. $6 \times 0 = ?$

14. What fraction of the rectangle is shaded?

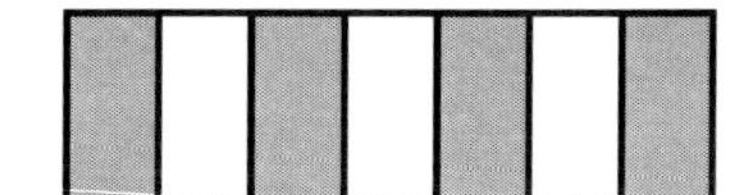

15. Which is greater, 5 quarters or 10 dimes?

16. Write *two-fifths* as a fraction.

17. $1{,}236 - 986 = ?$

18. How many ounces are in a pound?

19. What is the date of the second Thursday in April?

20. What is the date 2 weeks after April 10th?

April

S	M	T	W	T	F	S
				1	2	3
4	5	6	7	8	9	10
11	12	13	14	15	16	17
18	19	20	21	22	23	24
25	26	27	28	29	30	

1.	2.	3.	4.
5.	6.	7.	8.
9.	10.	11.	12.
13.	14.	15.	16.
17.	18.	19.	20.

Lesson #28

1. List the first 4 odd numbers.
2. $77 \times 4 = ?$
3. What time was it 15 minutes ago, if it is 12:30 now?
4. $5{,}657 + 9{,}821 = ?$
5. How many quarts are in 5 gallons?
6. $5 \times 5 = ?$
7. The answer to a subtraction problem is called the __________.
8. $2{,}883 - 977 = ?$
9. Draw perpendicular lines.
10. $244 \div 2 = ?$
11. Two figures with the same shape, but different sizes are _______.
12. Round *235,688* to the nearest hundred thousand.
13. Name the shape.

14. $81 \div 9 = ?$
15. Draw a heart. Draw a line of symmetry on the heart.
16. How many inches are in 2 feet?
17. $\frac{1}{4} + \frac{2}{4} = ?$
18. Mr. Talbot drove 88 miles on Monday, 123 miles on Tuesday, and 114 miles on Wednesday. How many miles did he drive altogether?
19. Does your Spelling book weigh about 1 pound or 1 ounce?
20. Draw a pentagon.

1.	2.	3.	4.
5.	6.	7.	8.
9.	10.	11.	12.
13.	14.	15.	16.
17.	18.	19.	20.

Lesson #29

1. $601 - 388 = ?$
2. Which digit is in the ten thousands place in *471,235*?
3. How many feet are in a yard?
4. $9{,}635 + 7{,}544 = ?$
5. Round *42,366* to the nearest hundred.
6. If it is 1:15 now, what time will it be in 15 minutes?
7. What is the mode of 12, 67, 44, 12, and 89?
8. $4 \times 9 = ?$
9. Name this shape.

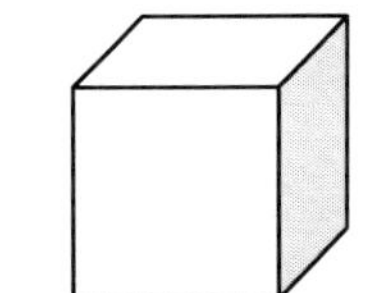

10. Is it more likely that Mario lives 2 feet from school or 2 miles from school?
11. $132 \div 3 = ?$
12. $3 \times 3 \times 3 = ?$
13. Is *3,507* an even or an odd number?
14. Draw a horizontal line.
15. $146 \times 6 = ?$
16. Any closed figure made up of line segments is called a _________.
17. Write *30,000 + 6,000 + 400 + 70 + 5* as a standard number.
18. Which has more sides, a hexagon or an octagon?
19. 36,874 ◯ 36,784
20. What fraction of the circle is shaded?

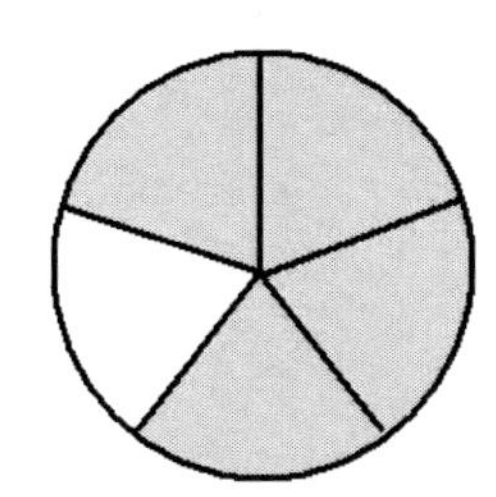

1.	2.	3.	4.
5.	6.	7.	8.
9.	10.	11.	12.
13.	14.	15.	16.
17.	18.	19.	20.

Lesson #30

1. List the even numbers between 41 and 49.
2. $3 \times 9 = ?$
3. Write the next number in the sequence. 25, 27, 29, …
4. $512 - 377 = ?$
5. Round *675* to the nearest ten.
6. Write *36,245* in expanded form.
7. Draw a line segment.
8. Write the time shown on this clock.

9. $556 + 897 = ?$
10. $68 \div 2 = ?$
11. How many ounces are in a pound?
12. $89 \times 3 = ?$
13. $36 \div 6 = ?$
14. What is the answer to a multiplication problem called?
15. Michael earns $9 a day. He works 3 days each week. How much money will Michael earn in 3 weeks?
16. Find $\frac{1}{3}$ of 12.
17. Are the two shaded fractions equivalent?

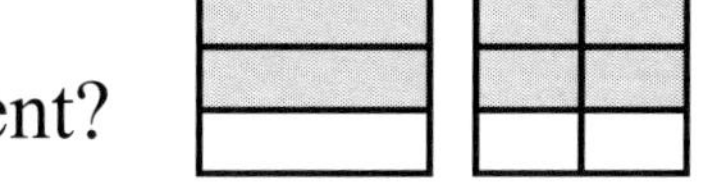

18. Draw 2 similar rectangles.
19. Does this show a line of symmetry?

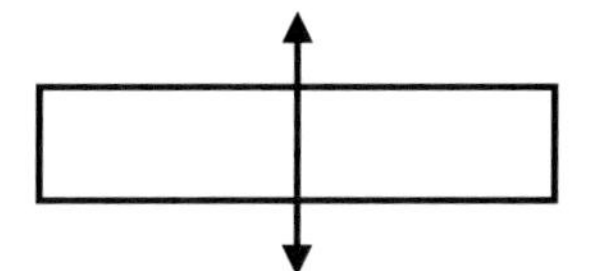

20. $(3 \times 2) \times 2 = ?$

1.	2.	3.	4.
5.	6.	7.	8.
9.	10.	11.	12.
13.	14.	15.	16.
17.	18.	19.	20.

Level 3

2nd Edition

Mathematics

Help Pages & "Who Knows?"

Help Pages

Vocabulary

Arithmetic Operations
Difference — the result or answer to a subtraction problem. Example: The difference of 5 and 1 is 4.
Product — the result or answer to a multiplication problem. Example: The product of 5 and 3 is 15.
Quotient — the result or answer to a division problem. Example: The quotient of 8 and 2 is 4.
Sum — the result or answer to an addition problem. Example: The sum of 5 and 2 is 7.

Geometry
Acute Angle — an angle measuring less than 90°.
Area — the size of a surface. Area is always given in square units ($feet^2$, $meters^2$,...).
Congruent — figures with the same shape and the same size.
Denominator — the bottom number of a fraction. Example: $\frac{1}{4}$ ➡ denominator is 4
Diameter — the widest distance across a circle. The diameter always passes through the center.
Fraction — a part of a whole. Example: ⊞ This box has 4 parts. 1 part is shaded. $\frac{1}{4}$
Line of Symmetry — a line along which a figure can be folded so that the two halves match exactly.
Numerator — the top number of a fraction. Example: $\frac{1}{4}$ ➡ numerator is 1
Obtuse Angle — an angle measuring more than 90°.
Perimeter — the distance around the outside of a polygon.
Radius — the distance from any point on the circle to the center. The radius is half of the diameter.
Remainder — the part left over when one number can't be divided exactly by another.
Right Angle — an angle measuring exactly 90°.
Similar — figures having the same shape, but different sizes.

Geometry — Polygons

Number of Sides	Name	Number of Sides	Name
3 △	Triangle	6 ⬡	Hexagon
4 □	Quadrilateral	8 ⯃	Octagon
5 ⬠	Pentagon		

Help Pages

Vocabulary

Measurement — Relationships	
Volume	**Distance**
3 teaspoons in a tablespoon	36 inches in a yard
2 cups in a pint	1760 yards in a mile
2 pints in a quart	5280 feet in a mile
4 quarts in a gallon	100 centimeters in a meter
Weight	1000 millimeters in a meter
16 ounces in a pound	**Temperature**
2000 pounds in a ton	0° Celsius - Freezing Point
Time	100° Celsius - Boiling Point
10 years in a decade	32° Fahrenheit - Freezing Point
100 years in a century	212° Fahrenheit - Boiling Point

Statistics

Mode — the number that occurs most often in a group of numbers. The mode is found by counting how many times each number occurs in the list. The number that occurs more than any other is the mode. Some groups of numbers have more than one mode.

Example: The mode of 77, (93), 85, (93), 77, 81, (93), and 71 is **93**.
(93 is the mode because it occurs more than the others.)

Place Value

Whole Numbers

2	7	1,	4	0	5
Hundred Thousands	Ten Thousands	Thousands	Hundreds	Tens	Ones

The number above is read: two hundred seventy-one thousand, four hundred five.

Help Pages

Solved Examples

Whole Numbers (continued)

When we **round numbers**, we are estimating them. This means we focus on a particular place value, and decide if that digit is closer to the next highest number (round up) or to the next lower number (keep the same). It might be helpful to look at the place-value chart on page 65.

Example: Round 347 to the tens place.

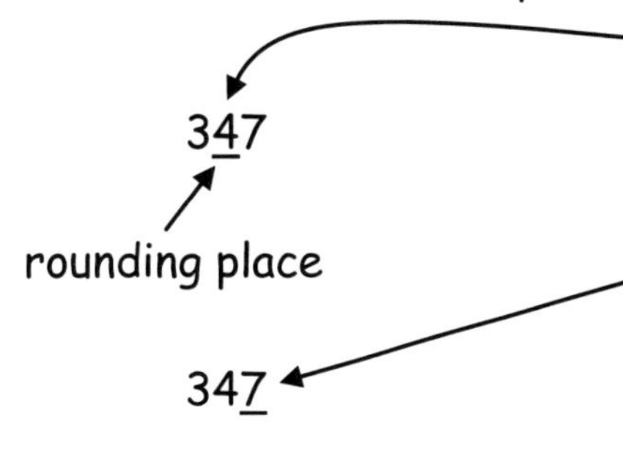

Since 7 is greater than 5, the rounding place is increased by 1.

350

1. Identify the place that you want to round to. What number is in that place? (4)
2. Look at the digit to its right. (7)
3. If this digit is 5 or greater, increase the number in the rounding place by 1. (round up) If the digit is less than 5, keep the number in the rounding place the same.
4. Replace all digits to the right of the rounding place with zeroes.

Here is another example of rounding whole numbers.

Example: Round 4,826 to the hundreds place.

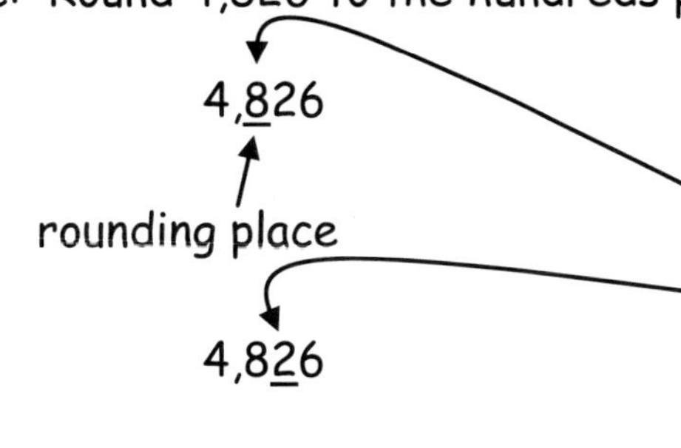

Since 2 is less than 5, the rounding place stays the same.

4,800

1. Identify the place that you want to round to. What number is in that place? (8)
2. Look at the digit to its right.
3. If this digit is 5 or greater, increase the number in the rounding place by 1. (round up) If the digit is less than 5, keep the number in the rounding place the same.
4. Replace all digits to the right of the rounding place with zeroes.

Help Pages

Solved Examples

Whole Numbers (continued)

When adding or subtracting whole numbers, first the numbers must be lined-up on the right. Starting with the ones place, add (or subtract) the numbers; when adding, if the answer has 2 digits, write the ones digit and regroup the tens digit (for subtraction, it may also be necessary to regroup first). Then, add (or subtract) the numbers in the tens place. Continue with the hundreds, etc.

Look at these examples of **addition**.

Examples: Find the sum of 314 and 12.

$$\begin{array}{r} 314 \\ +\ 12 \\ \hline 326 \end{array}$$

1. Line up the numbers on the right.
2. Beginning with the ones place, add. Regroup if necessary.
3. Repeat with the tens place.
4. Continue this process with the hundreds place, etc.

Add 6,478 and 1,843.

$$\begin{array}{r} {}^{1}\ {}^{1}\ {}^{1} \\ 6,478 \\ +1,843 \\ \hline 8,321 \end{array}$$

Use the following examples of **subtraction** to help you.

Examples: Subtract 37 from 93.

$$\begin{array}{r} {}^{8}\ {}^{13} \\ \not{9}\ \not{3} \\ -\ 3\ 7 \\ \hline 5\ 6 \end{array}$$

1. Begin with the ones place. Check to see if you need to regroup. Since 7 is larger than 3, you must regroup to 8 tens and 13 ones.
2. Now look at the tens place. Check to see if you need to regroup. Since 3 is less than 8, you do not need to regroup.
3. Subtract each place value beginning with the ones.

Find the difference of 425 and 233.

$$\begin{array}{r} {}^{3}\ {}^{12}\ \ \\ \not{4}\ \not{2}\ 5 \\ -\ 2\ 3\ 3 \\ \hline 1\ 9\ 2 \end{array}$$

1. Begin with the ones place. Check to see if you need to regroup. Since 3 is less than 5, you do not need to regroup.
2. Now look at the tens place. Check to see if you need to regroup. Since 3 is larger than 2, you must regroup to 3 hundreds and 12 tens.
3. Now look at the hundreds place. Check to see if you need to regroup. Since 2 is less than 3, you are ready to subtract.
4. Subtract each place value beginning with the ones.

Help Pages

Solved Examples

Whole Numbers (continued)

Sometimes when doing subtraction, you must **subtract from zero**. This always requires regrouping. Use the examples below to help you.

Examples: Subtract 261 from 500.

$$\begin{array}{r} \overset{4}{\cancel{5}}\,\overset{\overset{9}{\cancel{10}}}{\cancel{0}}\,\overset{10}{\cancel{0}} \\ -\,2\ 6\ 1 \\ \hline 2\ 3\ 9 \end{array}$$

1. Begin with the ones place. Since 1 is less than 0, you must regroup. You must continue to the hundreds place, and then begin regrouping.
2. Regroup the hundreds place to 4 hundreds and 10 tens.
3. Then, regroup the tens place to 9 tens and 10 ones.
4. Finally, subtract each place value beginning with the ones.

Find the difference between 600 and 238.

$$\begin{array}{r} \overset{5}{\cancel{6}}\,\overset{\overset{9}{\cancel{10}}}{\cancel{0}}\,\overset{10}{\cancel{0}} \\ -\,2\ 3\ 8 \\ \hline 3\ 6\ 2 \end{array}$$

Multiplication is a quicker way to add groups of numbers. The sign (×) for multiplication is read "times." The answer to a multiplication problem is called the product. Use the examples below to help you understand multiplication.

Examples: 3 × 5 is read "three times five."

It means *3 groups of 5* or 5 + 5 + 5.

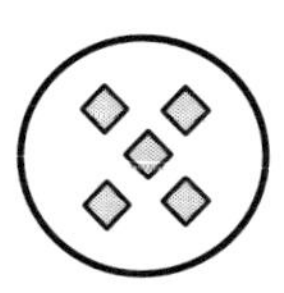
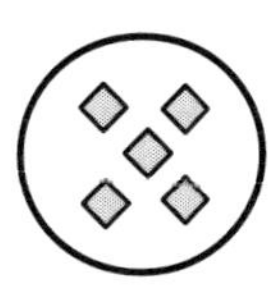
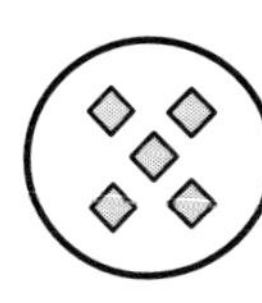

$3 \times 5 = 5 + 5 + 5 = 15$

The product of 3 × 5 is **15**.

4 × 7 is read "four times seven."

It means *4 groups of 7* or 7 + 7 + 7 + 7.

$4 \times 7 = 7 + 7 + 7 + 7 = 28$

The product of 4 × 7 is **28**.

Help Pages

Solved Examples

Whole Numbers (continued)

It is very important that you memorize your **multiplication facts**. This table will help you, but only until you memorize them!

To use this table, choose a number in the top gray box and multiply it by a number in the left gray box. Follow both with your finger (down and across) until they meet. The number in that box is the product.

An example is shown for you: $2 \times 3 = 6$

×	0	1	2	3	4	5	6	7	8	9	10
0	0	0	0	0	0	0	0	0	0	0	0
1	0	1	2	3	4	5	6	7	8	9	10
2	0	2	4	6	8	10	12	14	16	18	20
3	0	3	6	9	12	15	18	21	24	27	30
4	0	4	8	12	16	20	24	28	32	36	40
5	0	5	10	15	20	25	30	35	40	45	50
6	0	6	12	18	24	30	36	42	48	54	60
7	0	7	14	21	28	35	42	49	56	63	70
8	0	8	16	24	32	40	48	56	64	72	80
9	0	9	18	27	36	45	54	63	72	81	90
10	0	10	20	30	40	50	60	70	80	90	100

Help Pages

Solved Examples

Whole Numbers (continued)

When **multiplying multi-digit whole numbers**, it is important to know your multiplication facts. Follow the steps and the examples below.

Examples: Multiply 23 by 5.

$$\begin{array}{r} \overset{1}{2}3 \\ \times\ 5 \\ \hline 115 \end{array}$$

$3 \times 5 = 15$ ones or 1 ten and 5 ones

$2 \times 5 = 10$ tens + 1 ten (regrouped) or 11 tens

1. Line up the numbers on the right.
2. Multiply the digits in the ones place. Regroup if necessary.
3. Multiply the digits in the tens place. Add any regrouped tens.
4. Repeat step 3 for the hundreds place, etc.

Find the product of 314 and 3.

$$\begin{array}{r} 3\overset{1}{1}4 \\ \times\ \ 3 \\ \hline 942 \end{array}$$

$4 \times 3 = 12$ ones or 1 ten and 2 ones

$1 \times 3 = 3$ tens + 1 ten (regrouped) or 4 tens

$3 \times 3 = 9$ hundreds

Division is the opposite of multiplication. The symbols for division are ÷ and $\overline{)}$ and are read "divided by." The answer to a division problem is called the quotient. Remember that multiplication is a way of adding groups to get their total. Think of division as the reverse of this. In a division problem you already know the total and the number in each group. You want to know how many groups there are. Follow the examples below.

Examples: Find the quotient of 12 ÷ 3. (12 items divided into groups of 3)

The total number is 12. △ △ △ △ △ △ △ △ △ △ △ △

Each group contains 3. (△ △ △)(△ △ △)(△ △ △)(△ △ △)

How many groups are there? There are 4 groups.

$12 \div 3 = \mathbf{4}$

Divide 10 by 2. (10 items divided into groups of 2)

The total number is 10. △ △ △ △ △ △ △ △ △ △

Each group contains 2. (△ △)(△ △)(△ △)(△ △)(△ △)

How many groups are there? There are 5 groups.

$10 \div 2 = \mathbf{5}$

Help Pages

Solved Examples

Whole Numbers (continued)

Sometimes when you are dividing, there are items left over that do not make a whole group. These left-over items are called the **remainder**. When this happens, we say that "the whole cannot be divided evenly by that number."

Example: What is 16 divided by 5? (16 items divided into groups of 5)

The total number is 16.

Each group contains 5.

How many groups are there? There are 3 groups, but there is 1 left over.
The remainder is 1.

16 ÷ 5 = **3 R1** (This is read "3 remainder 1.")

The next group of examples involves **long division using one-digit divisors with remainders**. You already know how to divide single-digit numbers. This process helps you to be able to divide numbers with multiple digits.

Example: Divide 37 by 4.

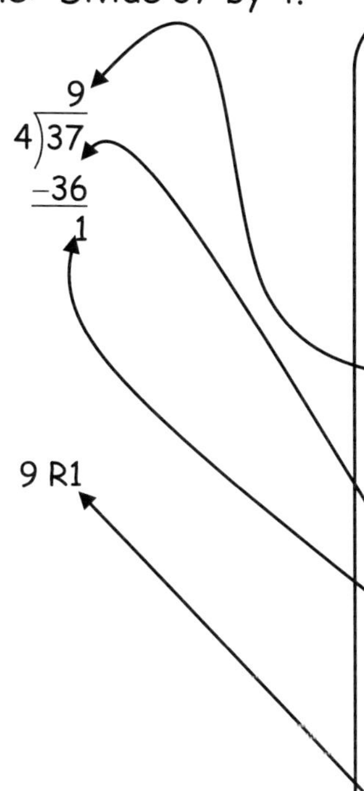

1. In this problem, 37 is the dividend and 4 is the divisor. You're going to look at each digit in the dividend, starting on the left.
2. Ask yourself if the divisor (4) goes into the left-most digit in the dividend (3). It doesn't, so keep going to the right.
3. Does the divisor (4) go into the two left-most digits (37)? It does. How many times does 4 go into 37? (9 times)
4. Multiply 4 × 9 (product = 36).
5. Subtract 36 from 37 (difference = 1). There's nothing left to bring down from above. Once this number is smaller than the divisor, it is called the remainder and the problem is finished. The remainder is 1.
6. Write the answer (above the top line) with the remainder. (9 R1)

Help Pages

Solved Examples

Whole Numbers (continued)

Example: What is 556 divided by 6?

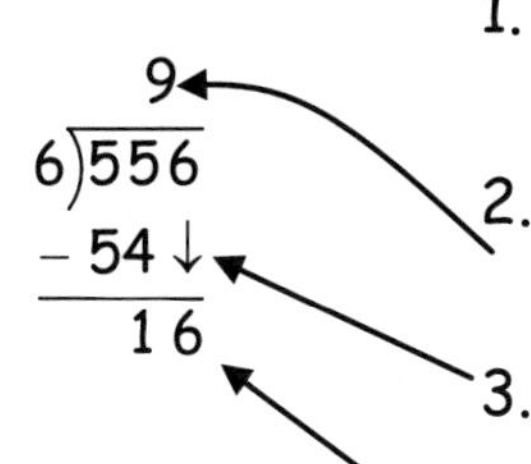

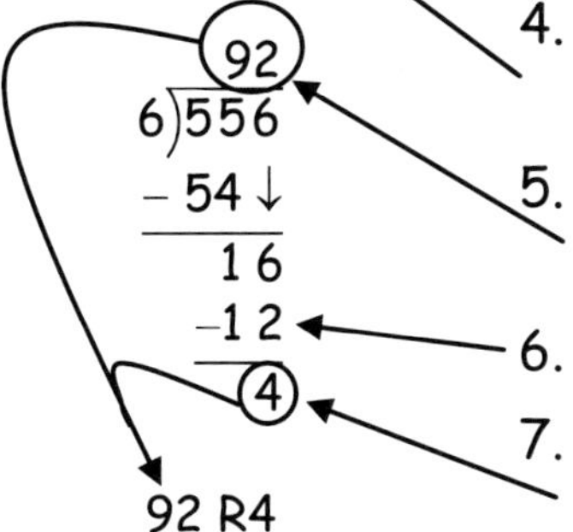

1. Ask yourself if the divisor (6) goes into the left-most digit in the dividend (5). It doesn't, so keep going to the right.
2. Does the divisor (6) go into the two left-most digits (55)? It does. How many times does 6 go into 55? (9 times)
3. Multiply 6 × 9 (product is 54).
4. Subtract 54 from 55. (1) Bring down the 6 ones from the first line. This leaves 16 left from the original 556.
5. Ask yourself if the divisor (6) goes into 16. It does. How many times does 6 go into 16? (2)
6. Multiply 6 × 2 (product is 12).
7. Subtract 12 from 16 (result is 4). There's nothing left to bring down from above. Once this number is smaller than the divisor, it is called the remainder and the problem is finished. The remainder is 4.
8. Write the answer with the remainder. (92 R 4)

Remember: The remainder can NEVER be larger than the divisor!

Fractions

A **fraction** is used to represent part of a whole. The top number in a fraction is called the **numerator** and represents the part. The bottom number in a fraction is called the **denominator** and represents the whole.

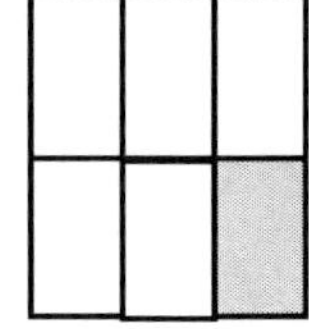

The whole rectangle has 6 sections.

Only 1 section is shaded.

This can be shown as the fraction $\frac{1}{6}$.

$\frac{1}{6}$ shaded part (numerator) / total parts (denominator)

To **add (or subtract) fractions with the same denominator**, simply add (or subtract) the numerators, keeping the same denominator.

Examples: $\frac{3}{5}+\frac{1}{5}=\frac{4}{5}$ $\frac{8}{9}-\frac{1}{9}=\frac{7}{9}$

Help Pages

Solved Examples

Decimals

Adding and subtracting decimals is very similar to adding or subtracting whole numbers. The main difference is that you have to line-up the decimal points in the numbers before you begin. Add zeros if necessary, so that all of the numbers have the same number of digits after the decimal point. Before you subtract, remember to check to see if you must regroup. When you're finished adding (or subtracting), bring the decimal straight down into your answer.

Example: Find the sum of 4.25 and 2.31.

$$\begin{array}{r} 4.25 \\ +\ 2.31 \\ \hline 6.56 \end{array}$$

1. Line up the decimal points. Add zeroes as needed.
2. Add (or subtract) the decimals.
3. Add (or subtract) the whole numbers.
4. Bring the decimal point straight down.

Example: Subtract 4.8 from 7.4.

$$\begin{array}{r} {}^{6}\not{7}.{}^{14}\not{4} \\ -\ 4.8 \\ \hline 2.6 \end{array}$$

Geometry

The **perimeter** of a polygon is the distance around the outside of the figure. To find the perimeter, add the lengths of the sides of the figure. Be sure to label your answer.

Perimeter = sum of the sides

Example: Find the perimeter of the rectangle below.

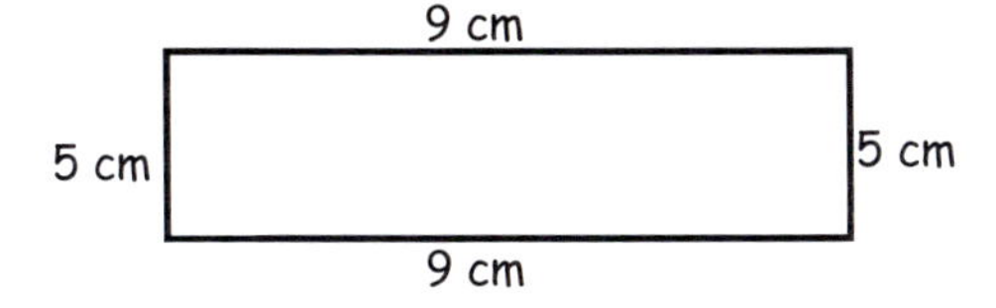

Perimeter = 5 cm + 9 cm + 5 cm + 9 cm
Perimeter = 28 cm

Example: Find the perimeter of the regular pentagon below.

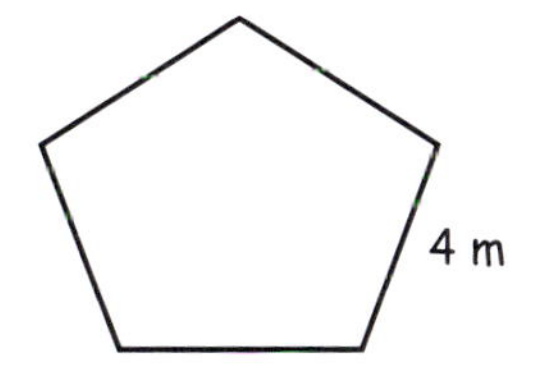

A pentagon has 5 sides. Each of the sides is 4 m long.

P = 4 m + 4 m + 4 m + 4 m + 4 m

P = 5 × 4 m

P = 20 m

Help Pages
Who Knows???

Sides in a Quadrilateral? .... (4)

Sides in a Pentagon? .... (5)

Sides in a Hexagon? .... (6)

Sides in an Octagon? .... (8)

Inches in a foot? .... (12)

Feet in a yard? .... (3)

Inches in a yard? ....(36)

Ounces in a pound? .... (16)

Pounds in a ton? ....(2000)

Cups in a pint? .... (2)

Pints in a quart? .... (2)

Quarts in a gallon? .... (4)

Years in a decade? .... (10)

Figures with the same size and shape? ....
....(congruent)

Figures with same shape, but different size? ....
.... (similar)

Answer to an addition problem? ....(sum)

Answer to a subtraction problem? .... (difference)

Answer to a multiplication problem? ...... (product)

Answer to a division problem? .... (quotient)

Level 3

2nd Edition

Mathematics

Answers to Lessons

Lesson #1		Lesson #2		Lesson #3	
1	1,625	1	27	1	
2	12 quarts	2	12 r 2	2	4:45
3	13	3	5	3	24
4	4	4	3,627	4	$3 \times 6 = 18$ $18 \div 3 = 6$ $6 \times 3 = 18$ $18 \div 6 = 3$
5	429	5		5	3,625
6	700	6	8 pages	6	
7	36	7	73	7	5 r 2
8	$8.44	8	congruent	8	108
9	656	9	63	9	
10	75¢	10	7,000	10	30
11	6	11	9,043	11	11:50 a.m.
12	4,000 + 200 + 10 + 6	12	3	12	$1.50
13	4:05	13	71	13	hexagon
14	hexagon	14	52, 54, 56	14	124
15	423	15	338	15	40 lbs.
16	odd number	16	10:30	16	90,000
17	<	17		17	12 months
18		18	1,296	18	554
19	100 pounds	19	8 cups	19	square
20	550 pounds of cans and bottles	20	No	20	(2, 4)

Lesson #4		Lesson #5		Lesson #6	
1	337	1	381	1	two thousand, five hundred forty-three
2	September	2	693	2	
3	111	3	addition	3	4,588
4	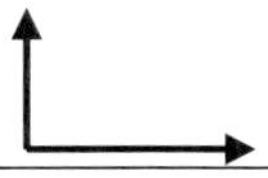	4	31, 33, 35, 37	4	$\frac{1}{2}$
5	13 r 2	5	13,090	5	3,300
6	8,341	6	16	6	<
7	100 cm	7	66¢	7	numerator
8	6	8	552	8	365 days
9	>	9	3	9	3,150
10	Yes	10	66,000	10	8 ft.
11	113	11	3 miles	11	124
12	7:30	12	7	12	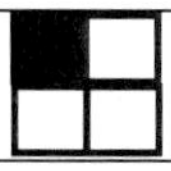
13	65	13	24 inches	13	
14	50,000 + 6,000 + 200 + 40 + 1	14	498	14	5
15	45	15	5,192; 5,219; 5,912	15	7
16	$\frac{3}{5}$	16	25	16	$4 \times 8 = 32$ $32 \div 4 = 8$ $8 \times 4 = 32$ $32 \div 8 = 4$
17	product	17	polygon	17	100
18	7 dimes	18		18	$2.25
19	numerator	19	$4.05	19	3:00
20	46 people	20	24 hours	20	230 more calories

Lesson #7		Lesson #8		Lesson #9	
1	6 inches	1	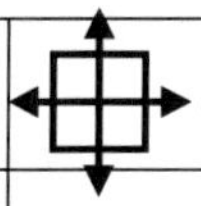 answers may vary	1	congruent
2	72,000	2	96	2	218
3	8 r 2	3	6 ft.	3	70
4	2,555	4	6:40	4	
5	cube	5	2,560	5	72 hours
6	difference	6	13,062	6	80
7	64,973	7	5	7	7
8	471	8	3 ft.	8	52¢
9	16 quarts	9	three-tenths	9	72
10	18	10	48	10	pentagon
11	No	11	No	11	2,616
12	5	12	$39.73	12	$\frac{3}{5}$
13	$\frac{2}{8}$	13	331	13	31
14	10:30	14	100 + 70 + 4	14	36
15	$\frac{4}{9}$	15	2 cups	15	sphere
16	22 books	16	6	16	11:10
17	282	17	odd number	17	$\frac{4}{7}$
18	odd number	18		18	denominator
19	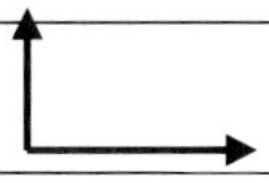	19	20 monkeys	19	108 cards
20	691	20	10 more	20	141

Lesson #10		Lesson #11		Lesson #12	
1	100 cm	**1**	5	**1**	9,590
2	658	**2**	1,347	**2**	2,000
3	octagon	**3**	9:28	**3**	5,280 ft.
4	39	**4**	72	**4**	300
5	$\frac{1}{4}$	**5**	quotient	**5**	cylinder
6	1,665	**6**	12,963	**6**	
7	8,000	**7**	rectangular prism	**7**	$\frac{1}{4}$
8	10 years	**8**	○ ○	**8**	4,189
9	□ □	**9**	248	**9**	7
10	2	**10**	3:50	**10**	9
11	508	**11**	740	**11**	100 cm
12	21 in.	**12**	4,567	**12**	35
13	842	**13**	9	**13**	1 ounce
14	8	**14**	240 seconds	**14**	numerator
15	3:30	**15**	576	**15**	1:15
16	19	**16**	18 ft. tall	**16**	114
17	pyramid	**17**	9	**17**	○ o
18	11 feet	**18**	12 quarters	**18**	$1.25
19	sum	**19**	7 r 3	**19**	Friday
20	3	**20**	10 students, 50 students	**20**	November 20th

Lesson #13		Lesson #14		Lesson #15	
1	polygon	1	4 × 5 = 20 20 ÷ 4 = 5 5 × 4 = 20 20 ÷ 5 = 4	1	426
2	4,755	2	1,227	2	pentagon
3	 •———•	3	400,000	3	12,822
4	6	4	8	4	multiplication
5	6:55	5	3,728	5	30
6	800	6	30,000 + 2,000 + 400 + 70 + 1	6	$1.10
7	259	7	11:55 am	7	50,000
8	tons	8	534	8	3 ft.
9	$\frac{1}{4}$	9	two-fifths	9	52
10	54	10	365 days	10	3 miles
11	1, 3, 5, 7, 9	11	4 pounds	11	1,575
12	7,537	12	64	12	6:25
13	127	13	$\frac{3}{4}$	13	64
14	300 minutes	14	<	14	8 cups
15	4	15	32	15	factors
16	zero	16	8 quarts	16	648
17	6 cups	17		17	Yes
18	9,865; 9,856; 9,685	18	8 r 1	18	9
19	74	19	similar	19	31, 35
20	22 ft.	20	$28	20	6 cupcakes

Lesson #16		Lesson #17		Lesson #18	
1	four thousand, three hundred fifty-three	1	37,528	1	$\frac{2}{5}$
2	$\frac{1}{5}$	2	6 r 2	2	329
3	36	3	21	3	8
4	1,234	4	even number	4	$2.80
5	1,048	5	90,000	5	460
6	49 days	6	36 in.	6	8 r 2
7	>	7	$38.62	7	3 ft.
8	324	8	8:30	8	polygon
9	6	9	216	9	>
10	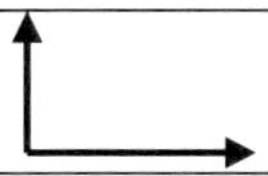	10	13,421	10	quotient
11	56¢	11	141 r 1	11	0, 2, 4, 6
12	$\frac{3}{4}$	12	$\frac{4}{7}$	12	4
13	addition	13	5,280 ft.	13	946,527
14	9:50	14	6	14	9
15	13,915	15		15	7:45
16	4	16	27	16	June
17	7	17	Yes	17	26, 38
18	3 lbs.	18	5,400	18	3
19	sphere	19	$\frac{7}{10}$	19	34 blocks
20	$\frac{3}{8}$	20	42, 44, 46, 48	20	18,424

Lesson #19		Lesson #20		Lesson #21	
1	7:30	**1**	686	**1**	3 quarters
2		**2**	800,000	**2**	40 minutes
3	16,663	**3**	1,300	**3**	$\frac{3}{8}$
4	218	**4**	8 cups	**4**	350,000
5		**5**	30	**5**	zero
6	$10	**6**	No	**6**	
7	544	**7**	95¢	**7**	16,458
8	1	**8**	323	**8**	42
9	9:30	**9**	6:45	**9**	10:00 a.m.
10	98¢	**10**	subtraction	**10**	16, 24
11	5	**11**	$5 \times 7 = 35$ $35 \div 5 = 7$ $7 \times 5 = 35$ $35 \div 7 = 5$	**11**	40
12	3,216	**12**	top	**12**	quadrilateral
13	7 ft.	**13**	5,026	**13**	5,000
14	30	**14**	12 years old	**14**	>
15	8	**15**	1 ounce	**15**	$\frac{2}{4}$
16	85, 87, 89	**16**		**16**	7
17	hexagon	**17**	36	**17**	rectangular prism
18	48	**18**	9	**18**	220
19	9 books	**19**	monkey	**19**	13
20	$\frac{2}{5}$	**20**	(1, 7)	**20**	20¢

Lesson #22		Lesson #23		Lesson #24	
1		**1**	9	**1**	14,603
2	83	**2**	30	**2**	\$15; \$24
3	9:20	**3**	2,239	**3**	24
4	7,682	**4**	$8 \times 9 = 72$ $72 \div 8 = 9$ $9 \times 8 = 72$ $72 \div 9 = 8$	**4**	9 r 2
5	4,665	**5**	4,675	**5**	161
6	51, 53, 55, 57	**6**	32 oz.	**6**	mode
7	12	**7**	9,000	**7**	6
8	53,940	**8**	4	**8**	9
9	924	**9**	forty-five hundredths	**9**	4,401
10	bottom	**10**	30,000 + 5,000 + 600 + 80 + 7	**10**	36 in.
11	28 quarters	**11**	$\frac{5}{8}$	**11**	
12	16 more	**12**	pyramid	**12**	200,000
13	7	**13**	64	**13**	1
14	28	**14**	8:30	**14**	\$19.75; 25¢
15	24 quarts	**15**	147	**15**	1:40
16	9 r 5	**16**	division	**16**	24,456
17	3	**17**	992	**17**	1 quarter, 2 dimes
18	100 cm	**18**	\$1.05	**18**	15
19	pounds	**19**	3,976; 5,319; 9,212	**19**	$\frac{4}{7}$
20	$\frac{3}{10}$, $\frac{7}{10}$	**20**	30 years	**20**	$\frac{2}{6}$

Lesson #25		Lesson #26		Lesson #27	
1	96	**1**	30,000	**1**	6
2	62, 64, 66, 68	**2**	12 months	**2**	2 hours
3	10,271	**3**	6:40	**3**	33
4	16 lbs.	**4**	octagon	**4**	$7
5	$\frac{1}{4}$	**5**	even number	**5**	9
6	63	**6**	90	**6**	sum
7		**7**	94	**7**	12:45
8	30 minutes	**8**	4:00	**8**	2,650
9	487	**9**	7,905	**9**	7,436
10	26 ft.	**10**	20,000 + 6,000 + 400 + 80 + 9	**10**	81, 83, 85, 87, 89
11	5,280 ft.	**11**	$1.50	**11**	48 inches
12	24	**12**	234	**12**	354,893
13	400 years	**13**	8	**13**	0
14	30,000	**14**		**14**	$\frac{4}{7}$
15	8 students	**15**	>	**15**	5 quarters
16	cylinder	**16**	8	**16**	$\frac{2}{5}$
17	32	**17**	multiplication	**17**	250
18	31	**18**	142 pages	**18**	16 oz.
19	5	**19**	85°F	**19**	April 8th
20	$\frac{4}{5}$	**20**	56	**20**	April 24th

Lesson #28		Lesson #29		Lesson #30	
1	1, 3, 5, 7	1	213	1	42, 44, 46, 48
2	308	2	7	2	27
3	12:15	3	3 ft.	3	31
4	15,478	4	17,179	4	135
5	20 quarts	5	42,400	5	680
6	25	6	1:30	6	30,000 + 6,000 + 200 + 40 + 5
7	difference	7	12	7	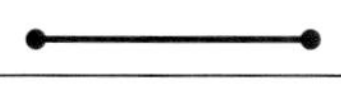
8	1,906	8	36	8	2:35
9	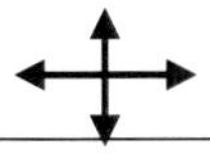	9	cube	9	1,453
10	122	10	2 miles	10	34
11	similar	11	44	11	16 oz.
12	200,000	12	27	12	267
13	rectangular prism	13	odd number	13	6
14	9	14		14	product
15		15	876	15	$81
16	24 in.	16	polygon	16	4
17	$\frac{3}{4}$	17	36,475	17	Yes
18	325 miles	18	octagon	18	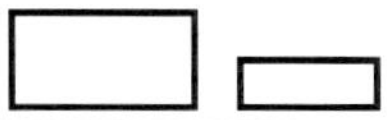
19	1 pound	19	>	19	Yes
20		20	$\frac{4}{5}$	20	12